Essentials of Hydroponics

Essentials of Hydroponics

Beatrice Pollard

MURPHY & MOORE
www.murphy-moorepublishing.com

Essentials of Hydroponics
Beatrice Pollard
ISBN: 978-1-63987-211-4 (Hardback)

MURPHY & MOORE
Published by Murphy & Moore Publishing,
1 Rockefeller Plaza,
New York City, NY 10020, USA

Cataloging-in-Publication Data

Essentials of hydroponics / Beatrice Pollard.
p. cm.
Includes bibliographical references and index.
ISBN 978-1-63987-211-4
1. Hydroponics. 2. Horticulture. 3. Plants--Nutrition. 4. Aquaponics. I. Pollard, Beatrice.
SB126.5 .E87 2022
631.585--dc23

For more information regarding Murphy & Moore Publishing and its products, please visit the publisher's website www.murphy-moorepublishing.com

Table of Contents

Preface

This book aims to help a broader range of students by exploring a wide variety of significant topics related to this discipline. It will help students in achieving a higher level of understanding of the subject and excel in their respective fields. This book would not have been possible without the unwavered support of my senior professors who took out the time to provide me feedback and help me with the process. I would also like to thank my family for their patience and support.

The method used to grow plants without the use of soil is known as hydroponics. Instead of using soil, it uses mineral nutrient solutions in a water solvent. This field is a branch of hydroculture. The field is emerging rapidly and is known to use less amount of water to grow plants. Cucumbers, tomatoes, peppers and lettuces are some of the plants that are grown using the various techniques of this field. Some of the techniques used in hydroponics are deep water, passive sub-irrigation, fogponics, continuous flow solution, static solution, aeroponics and ebb and flow sub-irrigation. Different mediums are suited for the use of different techniques. Hydroponics is an upcoming field of science that has undergone rapid development over the past few decades. Different approaches, evaluations, methodologies and studies related to this field have been included herein. In this book, constant effort has been made to make the understanding of the difficult concepts of this discipline as easy and informative as possible, for the readers.

A brief overview of the book contents is provided below:

Chapter - Introduction

Hydroponics is a sub-field of hydroculture that deals with growing of plants without soil but through mineral nutrient solutions in a water solvent. Two of the major hydroponic systems that are used widely are re-circulating system and drain to waste system. This is an introductory chapter which will briefly introduce about hydroponics.

Chapter - Methods and Techniques used in Hydroponics

There are various methods and techniques that are used in hydrophonics such as aeroponics, fogponics, anthroponics, aquaponics, rotary hydroponics, kratky method, vertical farming, etc. This chapter discusses the different methods and techniques used in hydroponics for a thorough understanding of the subject.

Chapter - Nutrient Solutions

The hydroponic nutrient solution is a liquid which consists of minerals in the raw water and nutrients added with fertilizers. Minerals such as calcium, magnesium, sulfur and trace elements such as boron, manganese, iron and zinc may be present in the source water. This chapter has been carefully written to provide an extensive understanding of nutrient solutions.

Chapter – Hydroponic Plants

Basil, rosemary, lettuce, spinach, strawberries, bell peppers, herbs, rose, lavender, green beans, tomatoes, etc. are some of the plants that grow best using hydroponic systems. This chapter has been carefully written to provide an easy understanding of these hydroponic plants.

Chapter – Equipment used in Hydroponic Farming

Different types of equipment are used in hydroponic farming for improving the quality of the hydroponic plants. Some of them are leaf sensor, spray nozzle, water chiller, net-pot, mesh pot, etc. The topics elaborated in this chapter will help in gaining a better perspective of the equipment used in hydroponic farming.

Beatrice Pollard

1 Introduction

Hydroponics is a sub-field of hydroculture that deals with growing of plants without soil but through mineral nutrient solutions in a water solvent. Two of the major hydroponic systems that are used widely are re-circulating system and drain to waste system. This is an introductory chapter which will briefly introduce about hydroponics.

Hydroponics

Plants grow through a process called photosynthesis, in which they use sunlight and a chemical inside their leaves called chlorophyll to convert carbon dioxide (a gas in the air) and water into glucose (a type of sugar) and oxygen. Write that out chemically and you get this equation:

$$6CO_2 + 6H_2O \rightarrow C_6H_{12}O_6 + 6O_2$$

There's no mention of "soil" anywhere in there and that's all the proof you need that plants can grow without it. What they do need is water and nutrients, both easily obtained from soil. But if they can get these things somewhere else say, by standing with their roots in a nutrient-rich solution they can do without soil altogether. That's the basic principle behind hydroponics. In theory, the word "hydroponics" means growing plants in water, but because you can grow plants without actually standing them in water, most people define the word to mean growing plants without using soil.

Although the benefits of hydroponics have sometimes been questioned, there seem to be many advantages in growing without soil. Some hydroponic growers have found they get yields many times greater when they switch from conventional methods. Because hydroponically grown plants dip their roots directly into nutrient-rich solutions, they get what they need much more easily than plants growing in soil, so they need much smaller root systems and can divert more energy into leaf and stem growth. With smaller roots, you can grow more plants in the same area and get more yield from the same amount of ground (which is particularly good news if you're growing in a limited area like a greenhouse or on a balcony or window-ledge inside). Hydroponic plants also grow faster. Many pests are carried in soil, so doing without it generally gives you a more hygienic growing system with fewer problems of disease. Since hydroponics is ideal for indoor growing, you can use it to grow plants all year round. Automated systems controlled by timers and computers make the whole thing a breeze.

It's not all good news; inevitably there are a few drawbacks. One is the cost of all the equipment you need containers, pumps, lights, nutrients, and so on. Another drawback is the ponic part of hydroponics: there's a certain amount of toil involved. With conventional growing, you can sometimes be quite cavalier about how you treat plants and, if weather and other conditions are on your side, your plants will still thrive. But hydroponics is more scientific and the plants are much more under your control. You need to check them constantly to make sure they're growing in exactly the conditions they need (though automated systems, such as lighting timers, make things quite a bit easier). Another difference (arguably less of a drawback) is that, because hydroponic plants have much smaller root systems, they can't always support themselves very well. Heavy fruiting plants may need quite elaborate forms of support.

There are various different ways of growing things hydroponically. In one popular method, you stand your plants in a plastic trough and let a nutrient solution trickle past their roots (with the help of gravity and a pump). That's called the nutrient-film technique: the nutrient is like a kind of liquid conveyor belt it's constantly sliding past the roots delivering to them the goodness they need. Alternatively, you can grow plants with their roots supported by a nutrient-enriched medium such as rockwool, sand, or vermiculite, which acts as a sterile substitute for soil. Another method is called aeroponics and it's typified by a popular product called the AeroGarden. Although the name suggests you're growing plants in air, the roots are actually suspended inside a container full of extremely humid air. Effectively, the roots grow in a nutrient-rich aerosol a bit like a cloud packed full of minerals.

In theory, you can grow any plant hydroponically but as is always the case with gardening some things inevitably do better than others. Fruit crops such as tomatoes and strawberries, and lettuces and herbs, are among plants that do particularly well.

There are different types of hydroponic systems, classified as follows:

Hydroponic Systems in Liquid Medium

These systems have not substrates for crop development, which is produced directly on the water-bearing systems by various plants as:

- Deep Flow Hydroponics: NGS.
- Floating systems: Floating shelves.
- Systems for water depth: NFT.

Hydroponic Substrate Systems

In these systems, we grow using inert substrates irrigated by drip irrigation, subirrigation, or exudation. The most common substrates are perlite, rockwool, coconut fiber and peat.

- Crops benches or grooves.
- Growing up in a bag.
- Growing in individual containers or channels.
- Growing surface (sanded).

Aeroponic Systems

Involves growing by systems where the root stay outdoors, in a container that kept it in the dark, where the nutrient solution is applied as an aerosol mist. The use of hydroponic systems is justified in cash crops that require high control of the production process, such as greenhouse cultivation of cucurbits, tomatoes, pepper and strawberry.

Drains in Hydroponics

Hydroponic systems require to specify a percentage of drainage, especially in hydroponic substrate systems, in order not to salinize the root environment especially for hydroponics saline waters use.

Depending on the use to be given to the drainage system are classified into:

Systems Solution Lost

The drain water in these systems is not reused in production under hydroponics. It is normally used to irrigate other low plots under traditional irrigation systems, as they have unused nutrients not usual in hydroponic system.

Advantages of Hydroponic System

- Provides the root system of a constant humidity level all the time, regardless of weather or crop growth stage time.
- Reduce the risks of excessive irrigation, as the root suffocation.
- Provides better planning and operation of the plantation.
- Optimize the cost of water and fertilizers.
- Ensures irrigation throughout the root zone.
- Significantly reduces the problems of diseases caused by soil pathogens.
- Increases yields and improved production quality.

Hydroponic System Parts

- Greenhouse.
- Head of automated irrigation and support system of plants.
- Pumping equipment.
- Containers for Concentrated solutions of nutrients.
- Ducting fertigation application.

- Irrigation Scheduler.
- Receiver of drainage or effluent.

Materials Used

Installation of PVC or PE irrigation and accessories needed. Depending on the type of hydroponic irrigation could have:

- Growing Tables.
- Culture Sacks.
- Particular culture systems (NGS, NFT).
- Supporting canals cropping system.
- Drain pans.

Advantages of Hydroponics

Plants can be grown almost anywhere: Hydroponic plants can be grown without soil and where little light is available, and can be grown with little space. This is particularly useful in an urban environment where space cost is at a premium.

Better control over plant growth: The operator of the hydroponic system can control nutrient delivery and water delivery to the plants, and can grow plants using a wider variety of methods of nutrient and water versus conventional soil based planting.

Eliminate some tasks of soil based systems: Hydroponic plants do not require pulling of weeds, fumigation, aeration of soil or application of pesticides like conventional planting does.

Conserve water and nutrients: Hydroponic systems can function in a closed loop, allowing for the recycling of water and nutrients. This also reduces the carbon footprint of operations.

"Transplanting Shock" is less than in soil systems: Hydroponic seedlings can be grown in a variety of propagation medium that can easily be transplanted into the hydroponic system, as opposed to being removed from the soil media. This also reduces propagation time over soil based growing.

Provide the highest yield per area: Hydroponic systems can grow plants all year, and are capable of going vertical. Hydroponics can provide considerably more yields annually than soil based planting.

Disadvantages of Hydroponics

Planting requires diligence: Although the time invested per day is less with hydroponic systems than conventional soil planting, hydroponic plants cannot be left unattended or they will die, as they are reliant on the operator for water and nutrients. If hydroponic plants die, they cannot be revived.

High capital cost: Although hydroponics has lower costs resource wise during operation, the start-up costs of a hydroponic system are considerably more than conventional soil based planting. Capital costs are high because there is a requirement to buy equipment such as fans, pumps, lights, solutions, timers, growing containers and other instrumentation.

Certain plants cannot be grown with hydroponics: Certain plants either take up too much space, or cannot thrive without soil and are unsuitable for hydroponics.

Requires technical knowledge: A hydroponic system requires extensive technical knowledge and training in order to operate.

Very reliant on mechanical systems: Hydroponic systems are very reliant on the various systems in the network, such as the water delivery system. If the watering system fails, it may be disastrous to the plants: the plants may wither up and die relatively quickly.

Longer return on investment: Commercial plants may be unattractive at first because of the high start-up costs. It may be hard to convince people to invest in this system because of the seemingly unknown rate of return.

Organic Hydroponics

Organic hydroponics is a method of hydroponic gardening where the nutrient solution is comprised entirely of organic compounds.

In all of the six types of hydroponic gardening, it is a basic requirement that a nutrient solution be mixed and circulated in order to feed the plants. A hydroponically grown plant receives all of its nutrients through this solution and not the medium it's grown in.

This nutrient solution is comprised of a combination of fertilizers that can be controlled and monitored for maximum plant growth.

In organic hydroponics, this nutrient solution is comprised of all-natural and organic fertilizers and micronutrients. This can be achieved through a variety of ways, including the use of a compost tea, which provides many of the micronutrients a plant requires.

In theory, because all of the ingredients of organic hydroponics are organic, the hydroponically grown fruits and vegetables are in fact, technically organic. However, according to some proponents, organics is not just a matter of ingredients, but an entire natural process in which a plant is grown. They would argue that by its very nature, hydroponics can never be considered "organic".

On a scientific level, if all of the nutrients included in the nutrient solution are derived purely from organic sources, then the fruit borne of this method of gardening is also organic, despite the fact that artificial lighting and other electronic components were used to produce the hydroponic crops.

There is a long-standing debate as to whether soilless hydroponic systems can ever be completely organic. For the most part, this debate is the result of disputed ideas about what organics actually means. Even in the commercial horticultural world, things are not clear-cut when it comes to what is and what isn't considered organic.

In many parts of the world, certified-organic systems must have soil as the cornerstone of their production.

While the reasoning behind whether or not hydroponics is organic is still under debate, hobby growers need not bother with the large-scale logic. Instead, they should decide what organics means to them and follow techniques to fulfill their own ideologies.

For some growers, producing organic and natural crops from an indoor garden simply means avoiding the use of toxic chemicals such as synthetic pesticides, fungicides and sterilization agents. These types of growers choose to focus more on natural approaches while fully embracing hydroponic methods.

For others, growing organically means incorporating the use of organic nutrients combined with beneficial microbial populations similar to the methods used by soil-based organic producers. Unfortunately, it's not just a simple case of switching from traditional, fertilizer-salt-based nutrients to organic ones since many early hydroponic systems were never designed to be used with organic compounds and many growers have run into major issues when trying this.

Traditional hydroponic nutrients are made from fertilizers such as calcium nitrate, potassium nitrate, monopotassium phosphate, iron chelate and many others that, when dissolved into water, dissociate into ions ready for immediate uptake by plants. This is what allows for such rapid and balanced plant growth.

The plants never have to starve or wait for nutrient ions to become available. However, calcium nitrate and many others used in traditional hydroponic systems are not considered organic, but synthetic or man-made, and are not part of an organic system.

Replacing highly effective and carefully calculated fertilizer salts with organic nutrient sources is not easy. Manufacturers of hydroponic fertilizer products go to great lengths to get the ideal parts per million of each nutrient ion in their products so plants grow as fast and balanced as possible.

With organic nutrient sources, it's impossible to be so precise, so mineral deficiencies within organic hydroponic systems are often an issue. Also, organic nutrients contain a great deal of carbon, which non-organic nutrient products do not provide. This carbon is an ideal source of food for microbes in the nutrient solution and root zone, feeding both beneficial and pathogenic fungi and bacteria.

If unwanted microbes begin feeding on the carbon from organic nutrient sources, things can get a little toxic, creating slimy nutrient solutions, anaerobic root conditions, diseases and even plant die back. To avoid these problems, organic growers just starting out should begin slowly and with systems known to have a better success rate.

Suitable Hydroponic Systems and Growing Mediums

Among the most successful systems for organic hydroponics are aquaponic systems. If the system is run with the correct fish-to-plant ratio and a good rate of mineralization carried out by specific bacteria, aquaponics is one of the easiest approaches to organics.

Novice growers should start with low-nutrient-demanding crops, such as lettuce, salad greens and fresh herbs, and gradually build up to fruiting plants as the system matures and higher rates of mineralization occur. Even though aquaponic systems provide good levels of nutrient ions from fish waste, high-nutrient-demanding crops may still need a little supplementation, especially extra trace elements from time to time.

Aside from aquaponics, many successful, organically certified commercial hydroponic systems incorporate the use of a natural growing substrate to support the plants and provide a high surface area for the large population of microbes needed to process (mineralize) the organic fertilizer sources.

Many of these mediums use coconut fiber as a base because it is a natural material amended with a range of other substrates such as high-quality composts, vermicast (worm castings), perlite, bark, peat, vermiculite and even rice hulls. There are also organic growing mixes on the market specifically designed for hydroponic systems.

It is important for the growing medium to be free-draining and highly aerated. It should ideally contain some already-mineralized nutrients (compost and vermicast), and be a source of naturally occurring beneficial microbes, or have them inoculated in before planting occurs. A good medium for organic production has proven to be a mix of coconut fiber, coarse perlite and high-quality, fully-processed vermicast (20%) to provide some immediately available nutrients and diverse microbes that will process the organic nutrient solution when it is applied.

New organic substrates incorporated into a hydroponic system should be conditioned before use fully moistened, inoculated with microbes, fed a small amount of organic nutrients and left to sit in a warm growing area for at least 1-2 weeks before planting out. This gives microbial communities time to establish and build up numbers before starting to break down organic compounds into plant-usable nutrient ions ready for the first plants to go in.

Once plants are in place, the irrigation system needs to be run a little differently than traditional, non-organic hydroponic systems. Drip irrigation at the root zone is advised. The root zone should never be over-saturated; only a small volume of leachate from the base of the growing containers is required.

A plant's root zone in organic systems is finely balanced between the requirements of the microbes and the root system itself. Both require high levels of oxygenation, and overwatering excludes much of the aeration in the pores of the growing medium.

Organic systems have a higher requirement for oxygenation in the root zone due to the heavy populations of microbes required. Many failures in organic systems are directly related to overwatering and poor aeration.

EC and pH Levels

Electrical conductivity (EC) and pH control is also different in organic hydroponic systems. Many organic nutrients don't conduct electricity, so EC readings may not be a true indication of the concentration of an organic solution. A garden's pH levels also tend to run higher in a healthy, organic system than many growers are used to maintaining in standard hydroponics.

Since pH-lowering acids commonly used in hydroponics such as nitric and phosphoric acids are not organic, pH is best left to stabilize on its own. Some organic nutrient products naturally have a high pH, so growers should try to select those that have a more suitable pH range for use in soilless systems.

Organic Nutrients

There are a few different approaches to obtaining and using organic nutrient sources in a hydroponic system. It can be difficult to get a balanced and suitably high ratio of all the essential minerals from organic sources alone, so experimentation with different products is helpful. There is a range of liquid organic nutrient concentrates on the market, as well as some fertilizer salts that are considered organic and naturally occurring to help boost growth where required.

Generally, sulphate trace elements, such as iron sulphate, copper sulphate, zinc sulphate, manganese sulphate and magnesium sulphate (Epsom salt), are allowed under organic production, so these can be used to help round out any deficiencies that may occur with organic nutrients.

An organic iron chelate can be made by mixing iron sulphate with citric acid powder to replace the synthetic iron chelate (EDTA or DTPA) commonly used in traditional hydroponics to ensure ongoing iron availability. Growers often need to blend two or more organic liquid fertilizers to provide sufficient and balanced nutrient levels for most hydroponic crops.

An organic base product (often a concentrated fish emulsion base or similar) blended with an organic liquid calcium is a good place to start. An organic nitrogen product may be required later on. Growers should aim to use products that have been designed for soilless systems wherever possible.

The main difficulty with running organic systems is obtaining sufficient amounts of nitrogen and calcium, which are required in large amounts by plants. Organic systems rely on microbes in the root zone to convert organic compounds into plant-available nitrogen sources and sometimes this process does not occur fast enough for uptake.

Calcium is difficult to obtain rapidly from organic nutrients as it relies on the breaking down of calcium-containing materials such as limestone. Growers who have hard water sources containing naturally occurring calcium have a major advantage in this case because this form of calcium is readily available for plant uptake.

It is possible to make an organic nutrient solution completely from raw materials rather than relying on commercially bottled products. While liquid bio-digesters that turn raw organic materials into usable plant nutrients have been used by some growers, the more reliable method for smaller systems is vermiculture (worm farming).

Vermiculture is a highly efficient way of processing high-mineral raw materials, such as manures, limestone, blood and bone, fish meal, seaweed meal, guano and others, into usable, mineralized hydroponic nutrient solutions that also provide the benefits of a diverse population of beneficial microbes.

The successful use of vermiculture to process organic fertilizers relies on two things. First, the vermicast must be processed to completion and then extracted into water for use in a hydroponic system. The extract or liquid draining from the vermicast system should not be used as a nutrient solution until the vermicast itself has been fully processed.

Many worm juices on the market are highly diluted and often not balanced enough to use as a stand-alone nutrient solution. Second, the quality of the raw materials going into the vermicast system will determine how balanced the final nutrient will be. High-mineral sources, such as fish, blood and bone meal are often dried and ground into products, while food scraps, weeds and vegetation contain only low levels of minerals and will not make a nutrient solution concentrated enough for most hydroponic systems.

One of the main problems with organic nutrients is concentration. Most organic liquid products are not as concentrated as standard, salt-based fertilizer formulations, so plants may become weak, stretched and more prone to disease.

Growers need to be aware of what underfed plants look like and boost nutrient concentrations as soon as these conditions are detected. Nutrient additives and boosters, such as humic and fulvic acid, are a good addition to organic systems as they help facilitate nutrient uptake and are generally considered organic.

Pest and Disease Control

Once growers have established a healthy root zone and are feeding plants a suitable organic blend of nutrients, the next step in organic production tends to be pest and disease control. Synthetic or chemical pesticides and fungicides are not considered organic, which eliminates some of the highly effective controls many hydroponic growers depend on.

With organic production, prevention of pest and disease problems becomes even more essential. The use of screening vents, double-door entries, close inspection of planting material entering the growing area, sticky indicator traps and careful monitoring of plant health are essential in preventing or catching infestations early on.

Among the organically allowable pest control products, many indoor gardeners are already familiar with neem oil and extracts, which are derived from the Indian neem tree. Many neem spray products are considered organic and help control a wide range of insect pests.

Beneficial and predator insects as part of an integrated pest management program are also widely used by organic growers, and there are a range of microbial spray products on the market such as BT for caterpillars.

Disease control can be little more complicated. Fortunately, biological controls like Beauveria spp. and other non-synthetic fungicides based on microbial species are available. Local hydro shops will have more information on all of these options.

Running a successful organic hydroponic system involves some trial and error. What works for one grower may not work for another. There is a lot more to the biological side of organic plant nutrition than we understand, much of it to do with microbial balance and populations in the root zone and nutrient solution. Overall, being successful in organic production is an exciting achievement and is one more growers are experimenting with.

Hydroponics and Environmental Clean-up

Water pollution refers to any chemical, physical or biological change in the quality of water that is detrimental to human, plant, or animal health. Water pollution affects all the major water bodies of the world such as lakes, rivers, oceans and groundwater. Polluted water is unfit for drinking and for other consumption processes. It may also be not suitable for agricultural and industrial use.

Types of Water Pollution

Toxic Substances

The greatest contributors to toxic pollution are oil spills, herbicides, pesticides, industrial compounds and heavy metals such as mercury (Hg), cadmium (Cd), chromium (Cr), nickel (Ni), lead (Pb) arsenic (As), copper (Cu), zinc (Zn), among others. Organic pollution occurs when an excess of organic matter, such as manure or sewage, enters the water. When organic matter increases in a water body, the number of decomposers will increase. These microorganisms grow rapidly and use a great deal of oxygen during their growth. This leads to a depletion of oxygen as the decomposition process occurs and consequently limits oxygen availability to aquatic organisms.

As the aquatic organisms die, they are broken down by decomposers which lead to further depletion of the oxygen levels. A type of organic pollution can occur when inorganic pollutants such as nitrogen and phosphates accumulate in aquatic ecosystems. High levels of these nutrients cause an overgrowth of plants and algae. The enormous decay of this algae and plant matter become organic material in the water and lowers the oxygen level causing suffocation of fish and other organism in a water body. This overall process is known as eutrophication.

Thermal Pollution

Thermal pollution can occur when water is used as a coolant near a power or industrial plant and then is returned to the aquatic environment at a higher/cooler temperature than it was originally. Thermal pollution can have a disastrous effect on life in an aquatic ecosystem as temperature increases/decrease the amount of oxygen in the water, thereby reducing the aquatic life presence.

Natural Pollution

Also called Ecological pollution, takes place when chemical pollution, organic pollution or thermal pollution is caused by nature rather than by human activity. An example of natural pollution would be an increased rate of siltation of a waterway after a landslide which would increase the amount of sediments in runoff water. Another example would be when a large animal, such as a deer, drowns in a flood and a large amount of organic material is added to the water as a result. Major geological events such as a volcano eruption might also be sources of ecological pollution.

Sources of Water Pollution

The most important sources of water pollution are domestic wastes, industrial effluents and agricultural wastes. Other sources include oil spills, atmospheric deposition, marine dumping, radioactive waste and eutrophication.

- Domestic sewage: is wastewater generated from the household activities. It contains organic and inorganic materials such as phosphates, nitrates, heavy metal-containing wastes. Organic materials are food and vegetable waste, whereas inorganic materials come from soaps and detergents.
- Industrial Effluents: Manufacturing and processing industry wastes contain organic pollutants and other toxic chemicals. Some of the pollutants from industrial source include Cd, Pb, Hg, As, asbestos, nitrates, phosphates, oils, etc. Wastewater from food and chemical processing industries contribute more to water pollution than the other industries such as distilleries, leather processing industries and thermal power plants. Also dye industries generate wastewater which changes the water quality especially water color. Many of the big industries have come up with wastewater treatment plants.

  However, it is not the case with small-scale industries. Water can also become contaminated with toxic or radioactive materials from industry, mine sites and abandoned hazardous waste sites. For instance, in 1932, the Minamata disease in which nearly 1,800 people died and many more suffered occurred due to consumption of fish containing high amounts of methyl mercury. It was caused by release of methyl mercury from Chisso Corporation's chemical factory.

  Additionally, an indirect effect by industrial activity is when acid precipitation is caused as burning fossil fuels emit sulfur dioxide into the atmosphere. The sulfur dioxide reacts with the water in the atmosphere, creating rainfall which contains sulfuric acid. As acid precipitation falls into lakes, streams and ponds it can lower the overall pH of the waterway, affecting plant life, and subsequently the whole food chain. It can also leach heavy metals from the soil into the water, killing fish and other aquatic organisms. Because of this, air pollution is potentially one of the most threatening forms of pollution to aquatic ecosystems.
- Agricultural Waste: Includes manure, slurries and runoffs. Farms often use large amounts of herbicides and pesticides, both of which are toxic pollutants. These substances are particularly dangerous to life in rivers, streams and lakes, where toxic substances can build up over a period of time. The runoffs from these agricultural fields cause water pollution to the nearby water sources. The seepage of fertilizers and pesticides causes groundwater pollution, which is commonly known as leaching. Fertilizers can increase the amounts of nitrates and phosphates in the water, which can lead to eutrophication. Allowing livestock to graze near water sources often results in organic waste products being washed into the waterways and can also lead to eutrophication.

Heavy Metal Pollution in Water

Heavy metal pollution of freshwater environments is a serious environmental problem in the industrial areas. Water pollution by heavy metals (elements with an atomic density greater than 6 g cm-3) has become therefore a global issue that need considerable attention towards combating. The common heavy metals that have been identified in polluted water include As, Cu, Cd, Pb, Cr, Ni, Hg and Zn. The release of these metals without proper treatment poses a significant threat to public health because of their persistence, biomagnification and accumulation in food chain. Their presence in water is due to discharges from residential dwellings, groundwater infiltration and industrial discharges. Their occurrence and accumulation in the environment is a result of direct or indirect human activities, such as rapid industrialization, urbanization and anthropogenic sources.

Severe toxic effects of heavy metal intake include reduced growth and development, cancer, organ damage, nervous system damage, and in extreme cases, death. The danger of heavy metal pollutants in water lies in two aspects of their impact. Firstly, heavy metals have the ability to persist in natural ecosystems for an extended period. Secondly, they have the ability to accumulate in successive levels of the biological chain, thereby causing acute and chronic diseases.

Water Treatment for Heavy Metal Removal

Several methods of removing heavy metals from water based on chemical and microbiological processes have been developed with a degree of success. Control over the quality and composition of industrial waste, including the removal of heavy metals, may take advantage principally of physicochemical methods based on chemical precipitation and coagulation (flocculation) followed by sedimentation, flotation, ionic exchange, reverse osmosis, extraction, microfiltration, adsorption on activated carbon, etc. However, these techniques are associated with high costs if large volumes, low metal concentrations, and high clean-up standards are involved. The insufficient effectiveness of the traditional heavy metal removal from water techniques has led to the search for more economical and simple procedures for the primary and (or) final removal of heavy metals from wastewater. Among these promising techniques is the phytoremediation of industrial wastewater, which involves the removal of heavy metals by adsorption, accumulation, or precipitation using higher aquatic and terrestrial plants, and the subsequent processing, utilization or burial of the contaminated biomass in special areas.

Phytoremediation

The term phytoremediation ("phyto" meaning plant, and the Latin suffix "remedium" meaning to restore) actually refers to a diverse collection of emerging plant-based technologies that use either naturally occurring or genetically engineered plants for cleaning contaminated environments. The primary motivation behind the development of phytoremediative technologies is the potential for low-cost remediation. Phytoremediation use plants to remove, reduce, degrade, or immobilize environmental contaminants, primarily those of anthropogenic origin, with the objective of restoring area sites to functional conditions for private or public applications. Research on phytoremediation has focused on the use of plants to: (1) accelerate degradation of organic contaminants, usually in concert with root rhizosphere microorganisms, or (2) remove/extract hazardous heavy metals from soils or water. Phytoremediation of contaminated sites is appealing because it is relatively inexpensive and aesthetically pleasing to the public compared to traditional remediation strategies.

Response of Plants to Metal Pollution

The general response of plants growing on a metal contaminated soil is categorized into the following:

- Hyperaccumulators: These are species of plants that absorb and concentrate high levels of heavy metals either in their roots, shoots and/or leaves. By definition, a hyperaccumulator must accumulate at least 100 mg g-1 (0.01 % dry weight), Cd, As and some other trace metals, 1,000 mg g-1 (0.1 dry weight). cobalt (Co), Cu, Cr, Ni and Pb and 10,000 mg g-1 (1 % dry weight). Hyperaccumulators take up high amounts of a toxic substance, usually a metal or metalloids in their shoots during normal growth and reproduction. Hyperaccumulators are found in 45 different families with the highest among the Brassicaceae. One such hyperaccumulator, Thlaspi caerulescens is a well-known Zn hyperaccumulator able to accumulate close to 30,000 and 10,000 mg kg-1 Zn and Cd respectively in the shoot dry matter without growth reduction.
- Metal Excluders (tolerant): This category of plant species can grow on soil with concentration of a particular elements that are toxic to most other plants by means of preventing metals from entering their aerial parts and so maintain constant metal concentration in the soil around their roots.
- Metal Indicators: In this plants, the extent of metal accumulation reflects metal concentration in the rhizospheric soil/water. Indicator species have been reported for mine prospecting studies to find new ore bodies. In general, shows poor control over metal uptake and transport processes.

Types of Phytoremediation

Depending on the underlying processes, polluted matrix, applicability, and nature of the contaminant, phytoremediation can be broadly categorized as:

Phytodegradation

Also called phytotransformation, is the breakdown of organic contaminants taken up by plants through metabolic processes within the plant, or the breakdown of contaminants surrounding the plant through the effect of compounds (such as enzymes) produced by the plants. Complex organic pollutants are degraded into simpler molecules and are incorporated into the plant tissues to help the plant grow faster. Plants contain enzymes (complex chemical proteins) that catalyse and accelerate chemical reactions. Some enzymes break down and convert ammunition wastes, others degrade chlorinated solvents such as trichloroethylene (TCE) and others degrade herbicides.

Phytostimulation

Also called rhizostimulation or plant-assisted bioremediation/degradation, is the breakdown of organic contaminants in the rhizosphere area (water/soil surrounding the roots of plants) through microbial activity and enhanced by the presence of plant roots. This process is generally slower than phytodegradation. Microorganisms (bacteria, fungi) consume and digest organic substances for nutrition and energy. In this way, certain microorganisms can digest organic pollutants such as hydrocarbons, pesticides or solvents and break them down into harmless products in a process called biodegradation. Natural substances released by the plant roots (exudates) sugars, acids and alcohols contain organic carbon that provides food, attract microorganisms to the rhizosphere and enhance their activity.

Phytovolatilization

Is the uptake and transpiration of a contaminant by a plant (e.g. poplars), with release of the contaminant (mainly organic) or a modified form of the contaminant to the atmosphere. Phytovolatilization occurs as growing trees and other plants take up water and the contaminants. Some of these contaminants can pass through the plants to the leaves and evaporate, or volatilize, into the atmosphere.

Phytoextraction

Also called phytoaccumulation, refers to the uptake of metals from soil by plant roots into above-ground portions of plants. Certain plants, i.e. hyperaccumulators, absorb unusually large amounts of metals in comparison to other plants. After the plants are allowed to grow for some time, they are harvested and either incinerated or composted to recycle the metals. This procedure may be repeated as necessary to bring soil contaminant levels down to allowable limits. If plants are incinerated, the ash must be disposed of in a hazardous waste landfill, but the volume of ash will be less than 10% of the volume that would be created if the contaminated soil itself were dug up for treatment. Metals such as Ni, Zn and Cu are the best candidates for removal by phytoextraction because the majority of the approximately 400 known plants that absorb unusually large amounts of metals have a high affinity for accumulating these metals.

Phytostabilization

Is the use of certain plant species to immobilize inorganic contaminants in the groundwater and soil through adsorption onto roots, absorption and accumulation by roots, or rhizosphere-mediated precipitation. This process is intended to reduce the mobility of the contaminant and prevent migration to the groundwater or atmosphere, with a consequent reduction of the pollutant bioavailability. This technique can be used to revegetate sites where natural vegetation is lacking due to high metal concentrations in surface soils. Metal-tolerant species can be used to restore vegetation to the sites, thereby decreasing the potential migration of contamination through wind erosion and transport of exposed surface soils and leaching of soil contamination to groundwater.

Phytofiltration

More commonly called rhizofiltration is the use of roots to uptake also store contaminants from an aqueous growth matrix.

Rhizofiltration

Because heavy metal pollution affects the quality of drinking water supply and wastewater discharge, great efforts have been made in the last two decades to reduce pollution in water resources to reach environmental sustainability (Goal 7 of the Millennium Development Goals). This Topic is therefore aimed at giving a general overview of rhizofiltration an hydroponic-based environmental biotechnology - as a cost-effective and sustainable alternative for the remediation of heavy metals pollutants in drinking water and wastewater treatment systems.

As mentioned previously, metal pollutants in wastewater, superficial water and groundwater are most commonly removed by chemical precipitation or flocculation, followed by sedimentation and disposal of the resulting sludge. A promising alternative to these conventional clean-up methods is rhizofiltration, a plant-based technique designed for the removal of metals in aquatic environments. In rhizofiltration plant roots grown in water absorb, concentrate and precipitate toxic metals and organic chemicals from polluted effluents. The plants can be used as filters in constructed wetlands or in a hydroponic setup.

Rhizofiltration Technology

Rhizofiltration is primarily used to remediate extracted groundwater, surface water, and wastewater with low contaminant concentrations. It is defined as the use of plants, both terrestrial and aquatic, to absorb, concentrate, and precipitate contaminants from polluted aqueous sources in their roots. Rhizofiltration can be used for Pb, Cd, Cu, Ni, Zn, and Cr, which are primarily retained within the roots. The plants to be used for rhizofiltration cleanup are raised in greenhouses with their roots in water. Contaminated water is either collected from a waste site or the plants are planted in the contaminated area allowing an insitu treatment, minimizing disturbance to the environment, where the roots then take up the water and the metal contaminants dissolved in it. As the roots become saturated with contaminants, they are harvested.

During this hydroponic process, plants can adsorb or precipitate onto roots (or absorb into the roots) the metal contaminants that are in solution surrounding the root zone. Changes in rhizosphere pH and root exudates may also cause metals to precipitate near root surfaces. As roots become saturated with the metal contaminants, plants can be harvested for disposal or reutilization. Dushenkov et al., suggested that that plants used for rhizofiltration should preferably accumulate metals only in the roots since the translocation of metals from roots to shoots would decrease the cost-effectiveness of rhizofiltration by increasing the amount of contaminated plant residue needing disposal. In contrast, Straczek et al. suggest that the capacity for rhizofiltration can be increased by using plants with an enhanced ability to translocate metals within the plant. Despite this difference in opinion, it is apparent that proper plant selection should be based in the total amount of metal removed from the polluted water indistinctly whether metal is accumulated in roots or shoots.

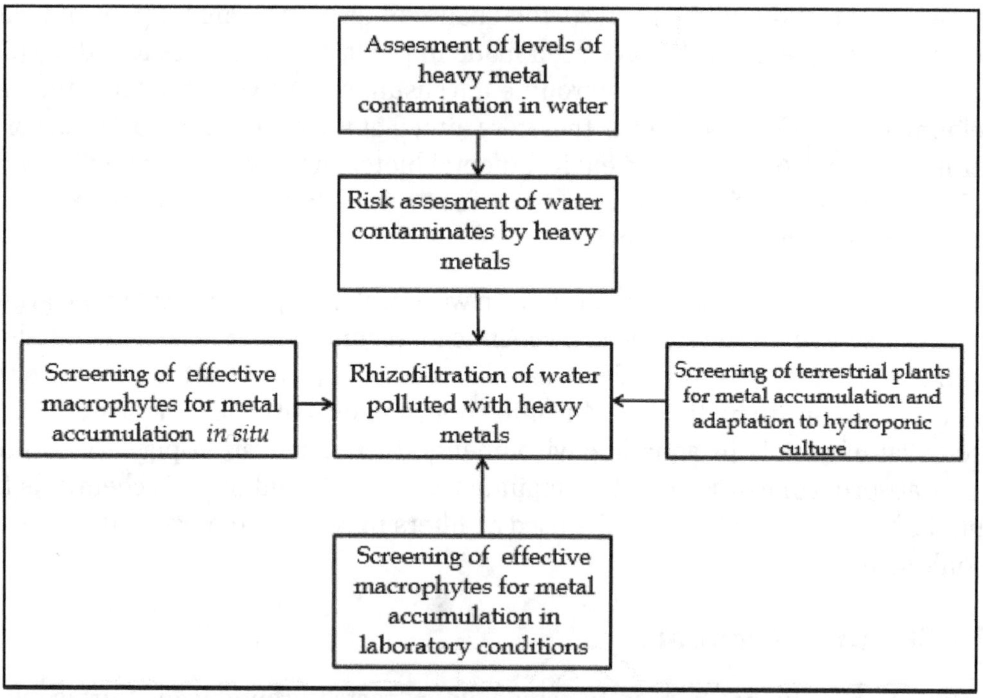

Phytoremediation process for water contaminated by heavy metals.

Plant Species for Rhizofiltration

Dushenkov & Kapulnik described the model characteristics of plants used for rhizofiltration. Plants should be able to tolerate and accumulate significant amounts of the target metals in conjunction with easy handling, low maintenance cost, and a minimum of secondary waste requiring disposal. It is also desirable plants to produce significant amounts of root biomass or root surface area and high evapotranspiration rates.

Several aquatic species have shown the capacity to remove heavy metals from water, for instance, water hyacinth, pennywort, and duckweed. However, these plants have shown limited potential for rhizofiltration, because they are not efficient at metal removal, a result of their small, slow-growing root system. These authors also point out that the high water content of aquatic plants make difficult their drying, composting, or incineration.

Despite limitations, Mahmood et al. indicated that water hyacinth is effective in removing trace elements in waste streams. For example, Mahmood et al. demonstrated that water hyacinth would remove silver from industrial wastewater for subsequent recovery with high efficiency in a fairly short time. The accumulation of some other heavy metals and trace elements in many species of wetland plants has also been demonstrated. Water hyacinth has been used successfully in wastewater treatment systems to improve the quality of water by reducing the levels of organic and inorganic nutrients, and readily reducing the level of heavy metals in acid mine drainage water.

Trace element removal by wetland vegetation can be greatly enhanced by selection of appropriate wetland plant species. The selection is based on the types of elements to be remediated, the geographic location, microclimate, hydrologic conditions, soil properties, and known accumulation capacities of the species. Knowledge of the capabilities of different wetland plant species to absorb and transport trace elements under different conditions is important to know. One such plant is the vascular aquatic plant water hyacinth which is commonly found in tropical and subtropical regions of the world. Water hyacinth is a fast growing, floating plant with a reasonably well-developed fibrous root system and large biomass and it adapts easily to various aquatic conditions.

Hydroponic system involves aeration and therefore is not limited to aquatic species; it often makes use of terrestrial species with large roots and good capacity to accumulate inorganics.

Rhizofiltration using Terrestrial Plants

The advantages associated with rhizofiltration are the ability to use both terrestrial and aquatic plants for either in situ or ex situ applications. Terrestrial plants are thought to be more suitable for rhizofiltration because they produce longer, more substantial, often fibrous root systems with large surface areas for metal sorption. Another advantage is that contaminants do not have to be translocated to the shoots. Sunflower (Helianthus annuus), Indian mustard (Brassica juncea), spinach (Spinacia oleracea), corn (Zea mays) and tobacco (Nicotiana tabacum) are among the most promising terrestrial candidates for metal removal in water. The roots of B. juncea are effective in the removal of Cd, Cr, Cu, Ni, Pb, and Zn, sunflower removes Pb, U, 137Cs, and 90Sr and tobacco removes Cd, Ni, Pb, Cr, Zn, Cu, Hg and As from hydroponic solutions. Similarly, López-Chuken & Young and López-Chuken et al. observed that roots of hydroponically grown terrestrial plants such as B. juncea and Z. mays effectively removed Cd from aqueous solutions. Candelario-Torres et al. shown that N. tabacum plants effectively remove toxic metals, such as Pb, Cd and Cr from polluted solutions.

Perhaps, the best example of a successful rhizofiltration remediation program occurred at Chernobyl, Ukraine, where, sunflowers were successfully used to remediate radioactive uranium from pond water. Tobacco has been shown to develop long and hairy root systems when grown in nutrient solution, which create an extremely high surface area.

Rhizofiltration

Arthur et al., published an remarkable and extensive review of the use of rhizofiltration technologies applied to removal and further recovery of metals. While the extensive review of literature reported by the authors, only a few representative examples will be provided here. Rhizofiltration of metal-contaminated water was early investigated by Schulman et al. They developed a screening method to look for mutants of Brassica juncea that had enhanced Cd or Pb accumulation capabilities. The authors found that cell-wall binding and precipitation are the primary mechanisms of Pb accumulation in plants, thus the authors concluded that the hyperaccumulating characteristic of the mutant was due to the increased cell wall per unit of root weight. The ability to remove and recover heavy metals, including Pb, Cr, Zn, Cu and Ni from aqueous solutions was shown in experiments with Medicago sativa (alfalfa). Optimum binding was in aqueous solutions at pH 5, and tests showed that binding to alfalfa shoots occurred within five minutes. Similar results were reported by López-Chuken & Young, where a rapid initial reduction of Cd concentration in a hydroponic solution was observed after an initial 3 hours of exposition to maize plants. This was thought to signify a rapid equilibration between the root system and Cd in solution rather than true absorption into roots. Alfalfa biomass is also reported as an effective species for recovering Au(III) from aqueous solutions. Recovery of valuable metals by plants is called "phytomining".

Rhizofiltration of uranium (U) by terrestrial plants has been investigated by Dushenkov et al. They found that certain sunflower species had a high affinity for U and could concentrate it from water into the roots. Phytofiltration of chromium-contaminated water has been studied, and it has been found that several plant species can uptake the toxic Cr(VI) species and reduce the pollutant to the nontoxic form, Cr(III). Water hyacinth supplied with Cr(VI) in nutrient culture accumulated Cr(III) in root and shoot tissues. Reduction to the nontoxic form appeared to occur in the fine lateral roots. Phytofiltration of As from potable water has being evaluated using the brake fern (Pteris vittata). This plant species is known to tolerate high concentrations of As.

While some plants may have desirable characteristics for rhizofiltration of toxic metals from the environment, it is critical to better understand the mechanisms behind these capabilities before they can be exploited to the fullest extent for phytoremediation programs. Recently, much research has being conducted to elucidate the chemical and physiological interactions involved in metal adsorption and accumulation.

To improve the potential of candidate plants for rhizofiltration, several research studies have been undertaken to select lines and improve plants by breeding. Recently, research using genetically modified (GM) plants has been spreading. Some transgenic plants are designed to be dwarf species. In this way, most of the biomass goes to leaves, and the rest to produce a short stem, similar approach as in the 1960s 'Green Revolution' for food production. Under this circumstances, when the native plants compete with them, do not allow them to develop.

However, according to Gressel and Al-Ahmad, there are some considerations of the use of GM plants in field studies: The presence of wild relatives in the same area; harvesting time; the possibility that the plant become a weed; the possibility of gene flux and plant surviving on its own. Additionally, consideration to leave a fallow zone of about 8-18 m. There is also recommended that no food production at site be allowed during the next growing season. In general, whether GM plants or not, a problem for species used for phytoextraction should be considered: The risk that they can displace native species.

Challenges of Hydroponic Phytotechnologies

Plant Development and Variability

One of the main problems encountered by rhizofiltration research at laboratory level is the large variability in heavy metals content measured between plants exposed to the same treatment solutions. López-Chuken made several attempts to minimise this variation. For example, growing a large excess of seedlings in order to select uniform plants prior to treatment application, increase nutrient solution volume and nutrient concentration to reduce minor variation in conditions during the exposure period, validate digestion and analysis procedures against standard reference materials. In addition, it was shown that for an hydroponic trial using Z. mays which showed apparently erratic trends with Cd treatments there was, nevertheless, a good correlation between root and shoot analysis for individual plants which seems to suggest that the variability in response observed lay with individual plants rather than being caused by a methodical source of error. In general, plant variability in the hydroponic trials followed the qualitative trend: maize > Indian mustard > tobacco.

To date, hydroponic experiments dealing heavy metal uptake, generally involve the use of plants at early steps of development, for example an experiment used young seedlings (wheat harvested 6 days after sowing) with short exposure times to the treated solution ranging between 0 and 200 minutes. In general, seeds from the Fam. Gramineae normally contain high nutrient reserves. Therefore, for short-term trials at early seedling stage, these plants may be still partially absorbing energy from the endosperm, and not completely reflecting nutrient conditions from solution. It is therefore recommended that hydroponic trials dealing with metal uptake, should preferably use plants as mature as possible, and with enough metal exposure time (≥ 10 days). This would allow the study of more mature plants and also minimised any short-term effects arising from the transition from nutrient to treatment solutions.

Berkelaar & Hale suggested the use of short-term metal accumulation experiments to maintain aseptic conditions and avoid biodegradation when using organic ligands as a reservoir of chelated metal in solution. However, it has been demonstrated an initial rapid sorption of metals from treatment solutions which does not necessarily reflect the normal uptake rate of the plants growing under steady state conditions in the treatment solutions. This effect was concluded to be a rapid approach to a pseudoequilibrium state between root surface sorption sites and nutrient solutions. Furthermore, even when aseptic conditions are not strictly controlled, little effect in metal speciation was observed due to the presence of organic matter in solution of long-term trials (6-8 weeks).

Can Rhizofiltration Effectiveness be Extrapolated to Soil Pollution?

One of the main advantages of the hydroponic-based experiments is that the solution chemistry in contact with plant root surfaces can be unequivocally designed and controlled during uptake trials. This represents a significant advantage over the uncertainties intrinsic to contaminated soil studies. Furthermore, (arguably) reliable morphological and chemical analysis of roots is only possible when using hydroponic growth media. However, the disadvantages of altered plant physiology and short exposure time in hydroponic studies are also well recognised. The choice of soil or hydroponic systems for the study of metal uptake must be dictated by the intention of the study but will always remain a compromise between the desire for control over experimental conditions and the unrealistic side effects of using any medium other than a naturally contaminated soil with a field sown plant allowed to follow its full span of physiological development. Chaney et al., suggested that the best media to test accumulator capacity by plants is the naturally contaminated soil (long-term), preferably in situ since it represents realistic conditions.

One problem associated with metal uptake trials with hydroponic systems, and other artificial media, is that such studies often seem to adopt very high ‘unrealistic’ metal concentrations. Recent research has shown that ordinary plants can even reach the metal hyperaccumulator “definition”, for example Cd >100 mg kg-1 under artificial media conditions.

In a set of soil-based and hydroponic experiments using maize and Indian mustard plants it has been shown that the ratio of Cd concentrations (Cdshoot:Cdroot) for different maize species ranged between 0.09 and 0.43 (mean = 0.24; SD = 0.11) throughout the trials. These results indicated that for all treatments, whether in soil or nutrient solution and even using different maize varieties (hybrid W23/L317; salt tolerant 2001-196-1; mays PI596543 and Cameron), Cd concentrations in roots were consistently larger than in shoots. However, when B. juncea plants were used, the ratio of concentrations (Cdshoot:Cdroot) showed large differences between soil (1.63 - 2.65; mean = 2.18; SD = 0.51) and nutrient solution (0.02 - 0.05; mean = 0.03; SD = 0.01) trials, despite using the same variety of Indian mustard (var. G32192). These results may suggest that perhaps not all plants are suitable for hydroponic experiments because plants show different physiological Cd uptake responses when growing in soil or dissimilar artificial media conditions. In the case of B. juncea the rapid root-to-shoot transfer observed for soil trials was virtually suspended in solution culture. Thus, under hydroponic conditions some factor (or a combination of factors) controlling Cd accumulation by roots and translocation to shoots may be affected.

Importance of Root Surface Area in Expressing Metal Uptake

Hydroponic trials also offer some advantages over soil experiments where there is a particular interest in the role of the root morphology as a control over metal uptake rates. Although the morphology of roots grown in nutrient solutions will differ from those generated in soils, the entire root can be extracted without physical damage or contamination from soil particulates.

López-Chuken & Young in a Cd rhizofiltration trial using Z. mays observed that the dataset was effectively 'normalised' when including root morphological parameters to express Cd uptake rates by plants. In all cases the root surface area (RSA) was the morphological characteristic that best explained changes in Cd uptake by plants. However, other measured root characteristics, (i.e. volume, length, root projected area) were so strongly correlated (R= 0.88-1.00) with the RSA that the differences between expressing Cd uptake rates using these parameters were minimal. Furthermore, Berkelaar & Hale, in a nutrient solution trial growing wheat, expressed Cd uptake rates per unit root length or 'number of tips' with similar efficiency to that of RSA. It would be hence interesting to include, as future work, the use of rooting hormones on single varieties to investigate the potential enhancement of metal uptake rates by plants. A recent research (in hydroponics) has shown that adding the growth phytohormone IIA (indole-3-acetic acid) in combination with EDTA, increased Pb accumulation in leaves by about 2800% and by 600%, as compared to Pb content in leaves of lucerne plants exposed to Pb alone and with Pb/EDTA, respectively.

Utilization of Phytoremediation By-products

One of the main concerns about the use of plants with high metal phytoaccumulation characteristics is their post-harvest disposal. Rhizofiltration technologies applied to metal removal generally involve repeated cropping of plants in contaminated water, until the metal concentration drops to acceptable level. The ability of the plants to account for the decrease in water metal concentrations as a function of metal uptake and biomass production plays an important role in achieving regulatory acceptance. Although this may sound simple, several factors make it challenging in the field. One of the difficulties for commercial implementation of rhizofiltration has been the disposal of contaminated plant material. After each cropping, the plant is removed from the site; leading to accumulation of huge amounts of hazardous biomass waste that needs to be stored or disposed appropriately so that it does not pose any risk to the environment.

Composting and compaction has been proposed as post-harvest biomass treatment by some authors however, leachates generated by composted or compacted biomass will need to be collected and treated appropriately. It has been also reported that plant material may be dried, burned and disposed of in landfill as ash. Another promising route to utilize biomass produces by phytoremediation in an integrated manner is through thermochemical conversion process. If rhizofiltration could be combined with biomass generation and its commercial utilization as an energy source, then it can be turned into profit making operation and the remaining ash can be used as bio-ore, the basic principle of phytomining.

Thermochemical energy conversion best suits the rhizofiltration biomass waste because it cannot be utilized in any other way as fodder and fertilizers. Combustion is a rudimentary method of burning the biomass and should be applied only under controlled conditions, whereby volume is reduced to 2–5 %. This method of plant matter disposal has to be carefully evaluated as burning the metal bearing hazardous waste in open will releases the gases and particulates to the environment.

Another alternative is the process called "gasification" through which biomass material can be subjected to series of chemical changes to yield clean and combustive gas at high thermal efficiencies. This mixture of gases called as producer gas and/or pyro-gas that can be combusted for generating thermal and electrical energy. The process of gasification of biomass in a gasifier is a complex phenomenon; it involves drying, heating, thermal decomposition (pyrolysis) and gasification, and combustion chemical reactions, which occurs simultaneously and it may be possible to recycle the metal residue from the ash.

Cost Estimates using Rhizofiltration

Rhizofiltration is a cost-competitive technology in the treatment of large volumes of water containing low, but environmental significant concentrations of heavy metals such as Cr, Pb, and Zn. The commercialization of rhizofiltration systems need to be driven by cost-effectiveness as well as by such technological advantages as applicability to many real conditions, ability to treat high volumes, lesser needed for chemicals, reduced volume of secondary waste, possibility of recycling and the almost secure likelihood of regulatory and public acceptance.

This hydroponic-based phytotechnology has worked effectively at test sites near the Chernobyl nuclear plant in Ukraine. It has been estimated that the cost to remove radionuclides from water using sunflowers would be between $2 and $6 per 1,000 gallons, including disposal costs On the other hand, a standard treatment of microfiltration and precipitation would cost nearly $80 per 1,000 gallons. Glass estimated that depending on the pollutant, substrate, and alternative remediation methods available, phytoremediation could be typically 2–10-fold cheaper than conventional remediation methods.

Despite obvious advantages, the application of this plant-based technology may be more challenging and susceptible to failure than other methods of similar cost. The production of hydroponically grown plants and the maintenance of successful hydroponic systems in the field will require the qualified personnel, and the facilities and specialized equipment required could exceed the original estimated cost. Perhaps the fundamental benefit of this remediation method is related to positive public perception. Using plants at a site where contamination exists conveys the idea of cleanliness in an area that would have normally been perceived as polluted.

Rhizofiltration and Sustainable Development

Contaminated water in the urban environments and rural areas represents a major environmental and human health problem in the world. Some plants possess pronounced capacity and ability for the metabolism and degradation of many contaminants and are regarded as "green livers" acting as a sink for environmentally harmful contaminants. It has been reported that green space programmes are conducted in most countries to check the increasing levels of carbon dioxide which causes global warming. But when properly managed and handled through the use of some plants that have phytoremediation property, green space would not only clean the atmosphere of its excess carbon dioxide but also the soil and water from its contaminants.

Developing cost effective and environmentally friendly technologies for the remediation of soils and wastewaters polluted polluted with toxic substances is a topic of global interest. The Millennium Development Goals-MDGs agreed by the international community includes "environmental basic sanitation" as a critical target. The necessity in decontaminating polluted sites is recognised worldwide, both socially and politically, because of the increasing importance placed on environmental protection and human health. Based on the success recorded by various studies on phytoremediation, rhizofiltration could represent a good alternative to contribute to achieve the above-mentioned goals.

Metals and other inorganic contaminants are among the most prevalent forms of contamination found at waste sites and the high cost of existing clean-up technologies has led to the search for new clean-up strategies that have the potential to be low-cost, lowimpact, visually benign, and environmentally sound. Rhizofiltration as an emerging new clean-up concept needs to be promoted and emphasized and expanded mainly in developing countries due to its low cost and potential to be applicable to a variety of contaminated sites. Selection of the appropriate plant species is a critical process for the success of this technology. Fast growing plants, adapted to hydroponic conditions, with high biomass and good metal uptake ability are needed.

An extra important advantage of rhizofiltration it that both terrestrial and aquatic plants can be used. Although terrestrial plants require physical support, they generally remove more contaminants than aquatic plants. This system can be either in situ (floating rafts on ponds) or ex situ (an engineered tank system). However, rhizofiltration has some intrinsic limitations pH of the polluted water has to be continually adjusted to obtain optimum metals uptake. b) chemical speciation and interaction of all metallic species in the influent has to be understood. An engineered system is required to control influent flow rate, d) plants may have to be grown in greenhouse, e) periodic harvesting and plant disposal are needed, and f) metal uptake results from laboratory studies might be overestimated and not be achievable under real conditions.

In general, phytoremediation is a multi- and inter-disciplinary technology that will benefit from research in many different areas. Much still remains to be discovered about the chemical and biological processes that underlie a plant's ability to detoxify and accumulate pollutants. Better knowledge of the biochemical mechanisms involved may lead to: a) the identification of novel genes and the subsequent development of transgenic plants with enhanced remediation capacities, and b) a better understanding of the ecological interactions involved (e.g. plant–microbe interactions in the rhizosphere) among others. This knowledge will help improve risk assessment during the design of rhizofiltration programs as well as alleviation of the associated risks during remediation. Adapting each rhizofiltration system to the specifics of polluted water will become more feasible as more information becomes available; for example to select a combination of plant species with different remediation capabilities to clean up sites containing a mix of contaminants. Preferentially, native plant species will be used in order to promote ecosystem restoration during the cleanup process.

An interesting perspective for phytoremediation could be the adoption of an integrated approach both for research and commercial purposes. Currently, most plant-based remediation research is carried out by scientists with expertise in only certain fields e.g. plant molecular biology, plant biochemistry, plant physiology, plant biochemistry, plant physiology, ecology, toxicology or microbiology but phytoremediation would be benefited more by a team of multidisciplinary researchers. Commercially to improve public acceptance phytoremediation could be integrated with landscape architecture with an attractive design so that the area may be used as a park or some other recreational place by the public after the remediation process.

Hydroponic Production Systems

Hydroponic systems for vegetable production are nowadays essential to maximize productions and increase yields.

Hydroponic systems help them in expanding their ability for a continuous production in a short growing period, require less space, and plants can be produced anywhere, i.e., in a small spaces with a controlled growth environment. Growers often reply that hydroponics always allows them to have higher productivities and yields without any constrains of climate and weather conditions. In addition, growers often claimed that quality of hydroponic produces is superior because it uses a highly controlled environment and enables a more homogeneous production without any loss of water and nutrients. Moreover, hydroponics is not dependent on seasonality, and therefore, their productivities are higher and homogenous throughout the year. Growers also often report that hydroponic productions are easier, and since they do not require cultural operations such as plowing, weeding, soil fertilization, and crop rotation, they are light and clean. However, the scientific evidences are often contradictory and different disadvantages are reported to justify their rejection: high initial costs, high technical and plant physiology knowledge, periodic work routines, and efficient electrical systems. It is also necessary and effective to control nutritional solutions and take daily measurements of liquid nutrients to avoid excess salinization and control microbial diseases and pests to avoid any loss of production. Nonetheless, growers often argue that this technique allows the possibility to grow healthier food and helps in the reduction of wastes. An example of this waste reduction can be seen in lettuce, the most hydroponically cultivated crop in the world, in which about 99% of their hydroponic leaves are valid and they can be sold to a value approximately of 10% more expensive than a lettuce grown traditionally. Moreover, with hydroponics, there is a better opportunity to place the fresh produces in the market since their average nutritional quality and consumer's acceptance are higher. In addition, growers reported that with hydroponics, some of the negative impacts of conventional agriculture are avoided including high and inefficient use of water, large land requirements, high concentrations of nutrients and pesticides, and soil degradation accompanied by erosion; issues that are much more in the nowadays concerns of consumers.

Worldwide consumers are increasingly interested in having more environment-friendly fresh vegetables due to the strong and well-established inverse relationship between vegetable consumption and the risk of many types of chronic and degenerative diseases like cancer, cardiovascular, and neurological disorders. Because of this growing consumer interest, the content of health-promoting compounds is becoming a vital consideration for fruit and vegetable growers. In fact, fresh vegetables and fruits are rich sources of bioactive compounds with significant health benefits, and these beneficial compounds can be influenced by several key factors including genotype selection and environmental conditions (light, temperature, humidity, atmospheric CO_2). Contrary to the conventional agricultural system, hydroponic relies on the manipulation of nutrients, which according to different authors allows having produces with high accumulation of some beneficial nutrients. However, questions about their safety are often raised.

There are considerable research studies regarding conventional and hydroponic production separately, but few have compared the impact of both on the nutritional quality of fresh vegetables.

Hydroponics and Quality of Vegetable Produces

Hydroponic production is the method of growing plants under soilless (i.e., soil less) conditions with nutrients, water, and an inert medium (gravel, sand, pearlite among others).

From the perspective of plant science, there are no differences between soilless and soil-grown plants, because in both systems, the nutrients must be dissolved in water before plants can absorb them. The differences reside in the way of nutrients that are available to the plants. In hydroponics, the nutrients are dissolved in water and the solution goes into the plant roots, which uptake the water with minerals toward different parts of plant. In the soil-based production, the elements stick to the soil particles, pass into the soil solution, where they are absorbed by the plant roots.

There are different types of hydroponics, depending on how they are characterized. One criterion is to classify as closed or open hydroponic systems. The hydroponic systems that do not use growing media are usually referred as closed systems, while hydroponic systems with growing media in a container may be closed or open depending on whether the nutrient solution is recirculated (closed) or is introduced on every irrigation cycle (open). In the closed systems, the nutrient concentrations are constantly recycled, monitored, and adjusted, while in open systems, the nutrient solution is discarded (but stored) after each nutrition cycle.

Another approach to classify hydroponic systems is to classify them based on the movement of the nutrient solution: active or passive. Active means that nutrient solution will be moved, usually by a pump, and passive relies on a wick or the anchor of the growing media. Others characterize the hydroponics with recovery or nonrecovery criteria. Recovery is when the nutrient solution will be reintroduced into the system, while nonrecovery means that nutrient solution is applied to the growing media and vanish after that.

Although there is a large diversity of criteria, there are three fundamental things for plants: (1) water/moisture, (2) nutrients, and (3) oxygen. All these different types of hydroponics must deliver those three important fundamental things to achieve success in plant production. Despite this diversity, the criteria most commonly used by growers, farmers, private companies, and researchers categorize hydroponic systems into six different types: Nutrient film technique (NFT), wick system, ebb and flow (flood and drain), water culture, drip system, and aeroponic system. Table summarizes the main characteristics of each system.

Hydroponic Systems and their Characteristics

Aeroponics: This is the most sophisticated and high-tech method, in which plants are suspended in special trails. Nutrients are sprayed, every few minutes, directly to the roots, which provide a light layer of nutrients. This system requires a regular monitorization of pumps to avoid any failure.

Deepwater culture (DWC): In a DWC system, a reservoir is used to hold the nutrient solution. The roots of plants are suspended the nutrient solution in order to get a constant supply of water, oxygen, and nutrients. In this system, an air pump is used to oxygenate the water, preventing the roots from drowning.

Drip system: In this system, the nutrient solution is set apart in a reservoir, and the plants are grown separately in a soilless medium. Drip systems dispense nutrients at a very slow rate, through nozzles, and the extra solutions can be collected and recirculated, or even allowed to drain out. With this system, it is possible to simultaneously grow several kinds of plants.

Ebb and flow: Also known as "flood and drain," is the less-commonly used system in hydroponics. This system utilizes a grow tray and a reservoir that is filled with a nutrient solution. A pump periodically floods the grow tray with nutrient solution, which then slowly drains away. In this system, plants are normally grown in mediums like rockwool or gravel, but if they need a substantial amount of moisture, this is substituted with vermiculite or coconut fiber due to their high capacity of excess moisture retention between floodings.

Nutrient film technique (NFT): Similar to aeroponics, the nutrient film technique (NFT) is the most popular hydroponic system. In this method, a nutrient solution is pumped constantly through channels in which plants are placed. When the nutrient solutions reach the end of the channel, they are sent back to the beginning of the system. This makes it a recirculating system, but unlike DWC, the plants roots are not completely submerged, which is the main reason for naming this method NFT.

Wick system: This is the easiest and simple method of hydroponics. It is a completely passive system, which means that nutrients are stored in a reservoir and moved into the root system by capillary action. With this system, we can find a diverse variety of growing medium such as perlite, vermiculite, coconut fiber, and other formulations.

The wick system is easy and inexpensive to set-up and maintain. The biggest drawback of this system resides in the poor oxygenation of plant roots and the large amount of nutrient solution that is required to reach efficiently to the plant root system.

Bioactive Compounds

According to Biesalski et al., it is widely accepted that bioactive compounds can be defined as essential and nonessential compounds that occur in nature as part of the food chain and with positive effect on human health. Bioactive compounds consist of chemicals found in small volumes in plants (leaves, roots, shoots, bark) and foods such as fruits, vegetables, nuts, oils, cereals, and grains. Bioactive compounds result from secondary metabolites of plants and are not essential for their daily functioning but play a significant role in the defense, attraction, signaling, and competition and thus are often named as secondary plant metabolites.

Types and Main Groups of Bioactive Compounds in Vegetables

Bioactive compounds may be classified according to different criteria. The most common classification is based on their pharmacological and toxicological effect. However, this is more relevant to the clinicians, pharmacist, or toxicologists and not for plant biologists, agronomists, or other researchers involved in plant-related studies. For these last groups, it is normal to classify them according to biochemical pathways and chemical classes. Table summarizes the main classes of bioactive compounds often found in plants and foods.

Table: Main classes of bioactive compounds commonly found in plants, classified according to chemical class criteria.

Chemical class	Health benefits	Example of primary food sources
Vitamins (C, E, and K)	Antioxidative protection of cells and membranes against reactive oxygen species (ROS), protection of lipoproteins, stabilization of cell membrane structures, modulation of immune response, inhibition of tumor growth in humans.	Fruits and vegetables.
Glycosides (cardiac glycosides, cyanogenic glycosides, glucosinolates, and anthraquinone glycosides)	Recent reports provide a large number of evidences about the chemopreventive role of different type of glycosides against cancers (bladder, prostate, esophagus, and stomach) and cancerous deleterious degenerative diseases. The glucosinolates, one of the most relevant group of glycoside compounds, are capable to activate enzymes involved in the detoxification of carcinogens, thus providing protection against oxidative damage.	Cruciferous vegetables, garlic, leek, onion.
Phenolics (natural monophenols and polyphenols including phenolic acids, flavonoids, aurones, chalconoids, flavonolignans, lignans, stilbenes, curcuminoids, and tannins)	The specific action of each phenolic is difficult to assess, because organisms absorb only a small part of it and in addition, they may suffer transformations. Epidemiological studies have claimed that polyphenols are capable of scavenging reactive oxygen species (ROS), inhibiting the peroxidation lipids from cellular membranes, preventing the LDL cholesterol from oxidation, and protecting DNA from mutations and oxidations.	Fruits and vegetables, tea, cocoa, wine, grapes, peanuts.

Carotenoids (carotenoids endowed with provitamin A activity, which are vital components of the human diet)	Although their bioactive mechanisms are still poorly understood, the heath importance of carotenoids is normally discussed in terms of antioxidant properties. Carotenoid-rich diets are correlated with a significant reduction in the risk of certain cancers, and heart and degenerative diseases. Epidemiological studies have shown that carotenoids may have anticancer and antimutagenic properties.	Green, orange, red, and yellow vegetables.
Plant sterols (phytosterols)	They regulate the fluidity and permeability of the phospholipid bilayers of plant membranes. Plant sterols have been hypothesized to have anticancer, antiatherosclerosis, anti-inflammation, and antioxidant activities.	Cruciferous vegetables, spinach, rice, soybean, wheat germ, wheat bran, nuts, and vegetable seeds.
Alkaloids (morphine, cocaine, solanine caffeine)	Morphine, cocaine, solanine, and caffeine are the most relevant alkaloids in plants. Besides the toxicity and their addictive effect, morphine and cocaine have been used in different formulations as anesthetics. The caffeine, the most known alkaloid and in high concentrations, is toxic and protects the plant seedlings from pests and plagues and prevents the germination of any other plants in the area (allopathic effects). In humans, caffeine has also been thought to reduce the risk of diabetes and heart disease, and recently, it has been associated with prevention of Alzheimer and Parkinson's diseases.	Papaveraceae (poppy family), coffee, cocoa, tea, Solanaceae (nightshade family), sweet peppers, chili peppers, jalapeno peppers eggplant, tomatoes, potatoes Atropa belladonna(deadly nightshade), Datura spp. (thorn apples).
Saponins	Among other properties, saponins are referred as having cardioprotective and hepatoprotective effects. Saponins have been observed to reduce blood cholesterol, stimulate the immune system, and inhibit the growth of cancer cells.	Beans, grains, chickpeas, oats, quinoa, asparagus, soy, red wine, and melons.

Hydroponic and Accumulation of Bioactive Compounds

Soil farmers experience these same types of variations with respect to soil health and fluctuations in environmental conditions. For example, water quality and variations in temperature and humidity can place stress on crops potentially changing their biochemical makeup regardless of the growing method being used. Because of these variations, studies to date comparing the nutritional content of produces grown hydroponically to soil grown have had mixed results, with some studies showing no difference between the two methods, while others showing that soilless systems fared either better or worse than soil-grown controls in the nutrient levels being tested. As you can imagine, experimental design and conditions vary widely between these studies and depend on how they were designed affecting the outcome and the significance of the findings.

Different studies have claimed that vegetables produced from hydroponics have better qualities than those from conventional soil-based cultivation. On the other hand, other studies have claimed that the exact differences between qualities of vegetables grown in soil or hydroponics are difficult to establish. Nonetheless, all authors in general seem to agree that hydroponic systems can be the best alternative when arable soil is scarce or their types are not ideal for the desired crop.

Although there are diverse and contradictory opinions, the general view of researcher seems to be that hydroponic can enhance the content of bioactive compounds. Recent studies have shown that in some high-value fresh vegetables, the hydroponic systems allow having higher nutritional quality due to high accumulation of bioactive compounds. Premuzic et al. found an increment of macro- and micronutrients as well as in antioxidants in hydroponic tomatoes, compared to soil-based production. Selma et al. found that hydroponic system was more effective in controlling microbial contamination as well as higher antioxidant compounds, since this method of production allowed a better maintenance of visual quality, control of browning, and more effective in controlling microbial contamination as compared with lettuces cultivated in soil. Pedneault et al. in Achillea millefolium found an accumulation of flavonoids in plants grown in hydroponic systems (0.43% dry weigh) compared to field-grown plants 0.38% dry weight). Also, Sgherri et al. found that hydroponic cultivation of basil (Ocimum basilicum cv. Genova) improved the antioxidant activity of both aqueous and lipid extracts, increasing the contents of vitamin C, vitamin E, lipoic acid, total phenols, and rosmarinic acid. Table presents some examples of studies about the accumulation of bioactive compounds under hydroponic systems.

Table: Comparative results between hydroponics and conventional soil-based production.

Crop	Results
Basil	The hydroponic cultivation improved the contents of vitamin C, vitamin E, lipoic acid, total phenols, and rosmarinic acid, as well as their antioxidant activities.
Lettuce	Hydroponics offered 11 ± 1.7 times higher yields compared to conventionally produced, but also required 82 ± 11 times more energy.
Lettuce	Levels of alpha-tocopherol here were higher in hydroponics compared to conventional soil-based production.
Lettuce	The content of lutein, beta-carotene, violaxanthin, and neoxanthin were lower in hydroponics compared to the soil-based production, due to less exposure of hydroponics to sunlight and temperatures, which had significant impact on carotenogenesis decreasing their levels.
Lettuce	Hydroponics-grown lettuce had significantly lower concentration of microorganisms compared to other in-soil-grown lettuce.
Onion	Total flavonoids were similar between hydroponics and soil-based cultivation.
Red paprika	The content of carotenoids capsorubin and capsanthin was higher in hydroponics (4.50 and 46.74 mg/100 g dry weight, respectively) compared to convention soil culture (2.81 and 29.57 mg/100 g dry weight, respectively).

Strawberry	Fruit yield per plant was 10% higher in hydroponic raspberries compared to soil-grown raspberries.
Sweet potato	Carotenes, ascorbic acid, thiamin, oxalic and tannic acids, and chymotrypsin and trypsin inhibitors were higher under hydroponics.
Tomato	No significant differences between hydroponic and nonhydroponic tomatoes in the levels of lycopene content (averaging 36.15 and 36.25 μg/g, respectively).
Tomato	Highly controlled conditions of electrical conductivity (EC) and salinity of water, pH, and nutrients provide optimum condition for enhancing the levels of sugars, Brix, pH, and organic acids, which are quality criteria of consumer acceptance toward tomato.

Based on the results obtained by the different authors, it seems evident that all of them reported as common reasons for the enhancement of bioactive compounds under hydroponics the tight control over the entire process of cultivation, particularly the amount and composition of nutrients and environmental conditions of temperature, humidity and light, and water salinity. Hydroponic operations, including the water recirculation systems, may provide ideal conditions for enactment of secondary metabolites, particularly when plants are placed under osmotic or salt stress, which boost the natural bioactive compounds of plants. Moreover, the saturation of light and temperatures by leaf receptors, often used in these systems, contributes to maximize photosynthesis and subsequent carbohydrate production that will be used for different biochemical mechanism including bioactive compounds biosynthesis, enhancing their content. For example, Greer and Weston found that in a controlled environment with lower temperatures, the content of anthocyanin in berries increased.

Other authors observed an increment of phenolic acids, flavonoids, and anthocyanins when the ratio of day/night temperature and day/night length was modified from low to high. Also, in a recent study with rocket salads (Eruca sativa, Eruca vesicaria, and Diplotaxis tenuifolia), it was found that setting long day lighting (16 h light, 8 h dark) at an intensity of 200 μmol m^{-2} s^{-1}, with a daytime temperatures of 20°C and night-time temperatures of 14°C, caused an enhancement in the content of polyphenols and glucosinolates. Likewise, day/night temperatures and day/night length, the nutrient solutions, the type of lights, and the levels of CO_2 can be used to enhance the content of bioactive compounds. Nutrient solutions with high electrical conductivity (EC) were shown to be efficient in the increment of lycopene (from 34 to 85%) in tomato cultivars. In a study with Gynura bicolor DC (a traditional vegetable from China and South East Asia) submitted to 80% red light and 20% blue light, supplemented with CO_2elevation from 450 (ambient reading) to 1200 μmol mol^{-1}, the content of bioactive compounds enhanced significantly. The content of anthocyanins and flavonoids rose from 400 to 700 mg 100 g^{-1} dry weight and from 250 to 350 mg 100 g^{-1} dry weight, respectively. Therefore, growing plants in a highly controlled environment might be an efficient alternative to maximize the production of bioactive compounds.

Although there are many advantages of the hydroponics compared to soil-based production, there are several aspects to be taken in account when we decide to choose hydroponics. These types of production systems require a regular irrigation and fertilization, which may otherwise result in possible contamination of surface and groundwater. In addition, they required an adequate management of pH, electrical conductivity (EC), dissolved oxygen, and temperature of nutrient solutions, because the ion concentrations may change with time, resulting in a nutrient imbalance. Therefore, real-time and periodical measurements of nutrient solutions are required, and adjustment of nutrient ratios is often required. In addition, to avoid infestations and diseases, disinfection systems are obligatory. All these aspects must be considered to achieve high quality without compromising the production yield and safety.

Automation and Robotics used in Hydroponic System

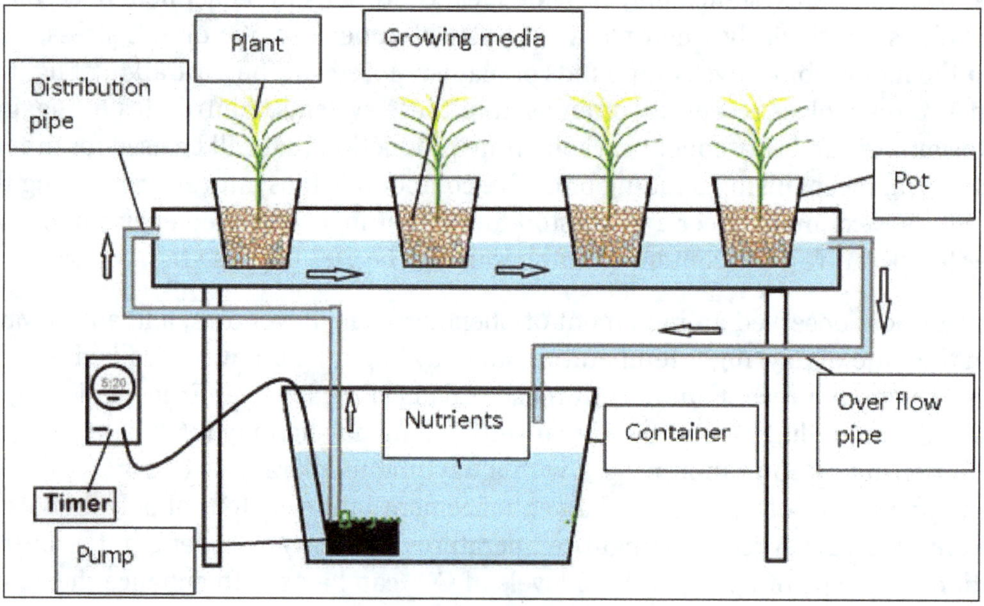

Nutrient film technique.

Hydroponic system requires periodic labor, a systematic approach, repetitive motion and a structured environment. Automation, robotics and IoT have allowed farmers to monitoring all the variables in plant, root zone and environment under hydroponics. This research introduces findings in design with real time operating systems based on microcontrollers; pH fuzzy logic control system for nutrient solution in embed and flow hydroponic culture; hydroponic system in combination with automated drip irrigation; expert system-based automation system; automated hydroponics nutrition plants systems; hydroponic management and monitoring system for an intelligent hydroponic system using internet of things and web technology; neural network-based fault detection in hydroponics; additional technologies implemented in hydroponic systems and robotics in hydroponic systems. The above advances will improve the efficiency of hydroponics to increase the quality and quantity of the produce and pose an opportunity for the growth of the hydroponics market in near future.

It is estimated that the total world population could reach 9.15 billion in 2050, and to increase the global food production, even more advances in agriculture must be made intensive in crop yields and in practices that are more friendly with the environment. Hydroponics is a method of growing plants in a water solution without soil. If the roots are suspended in a liquid medium or supported using an inert medium, the system is known as Nutrient Film Technique (NFT). In NFT systems, the plants (lettuce, leafy crops and herbs) are grown in channels (gullies) and fed continuously at a rate of approximately 1 L min^{-1}.

On the other hand, if the roots are floating (pool), the system is known as deep floating technique (DFT). The DFT systems are long, cement or wood rectangular reservoirs and lined with a durable polyliner. To keep the plants in net pots, holes are perforated in a foam board which rest on the surface of the water.

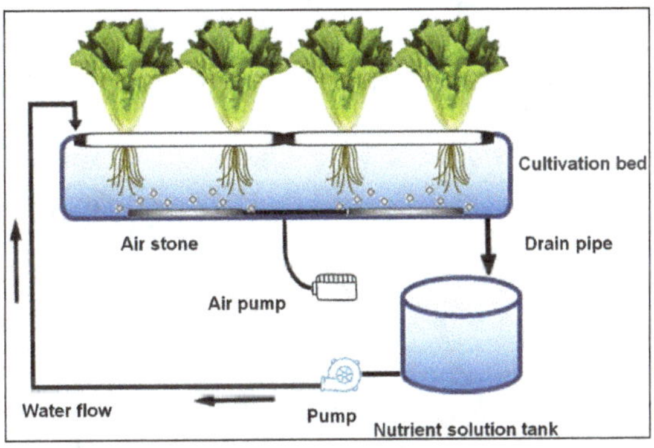

Deep floating technique.

Aeroponic systems are very similar to NFT systems, differing primarily in the spatial arrangement of cultivation channels. The cultivation space is optimized for the aeroponic plants are grown suspended in air, having as support PVC pipes which can be arranged horizontally or vertically, enabling a better exploitation of areas and installing a larger number of plants per square meter surface of the oven, obtaining thus a direct increase of productivity. Hydroponic systems, such as the deep flow technique, nutrient film technique or aeroponic systems, are essential tools in plant factories. To accomplish with this, hydroponic systems must collect a lot of information, since this allows a better diagnosis of the problems and better understand the development of hydroponic crops. Automatic sensors not only have the ones that can be read at predefined intervals, but also the readings of these sensors are stored so that higher results can be obtained for analysis and diagnosis resulting in higher crop yields and friendlier practices with the environment. These days, there are microcontrollers on the market that are compatible with a wide variety of sensors and can be used for automatic monitoring and robotics.

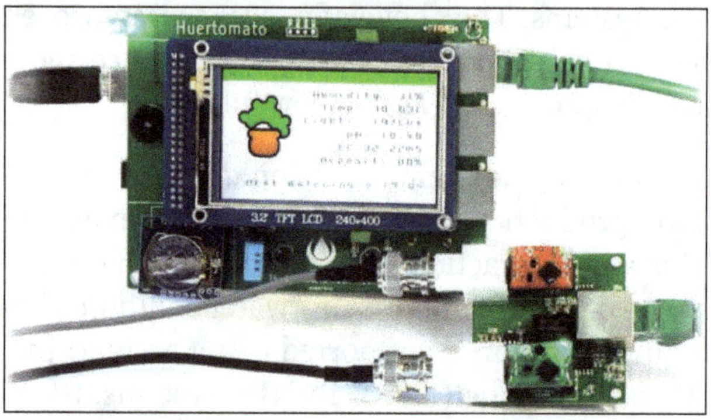

Huertomato microcontroller for measuring of humidity, water and air temperature, light, pH, electrical conductivity.

The emergence of Internet of Things (IoT) has allowed farmers to automate the hydroponic culture. Monitoring of water level, pH, temperature, flow and light intensity can be regulated using IoT, which allows for machine to machine interaction and controlling the hydroponic system autonomously and intelligently employing deep neural networks. The pH of the nutrient solution for most nutrient film technique is 6.0–7.0 for most plants grown in recirculating nutrient solution and 5.4–6.0 for substrate culture. There are also powerful computers that could store all this information and build a big database.

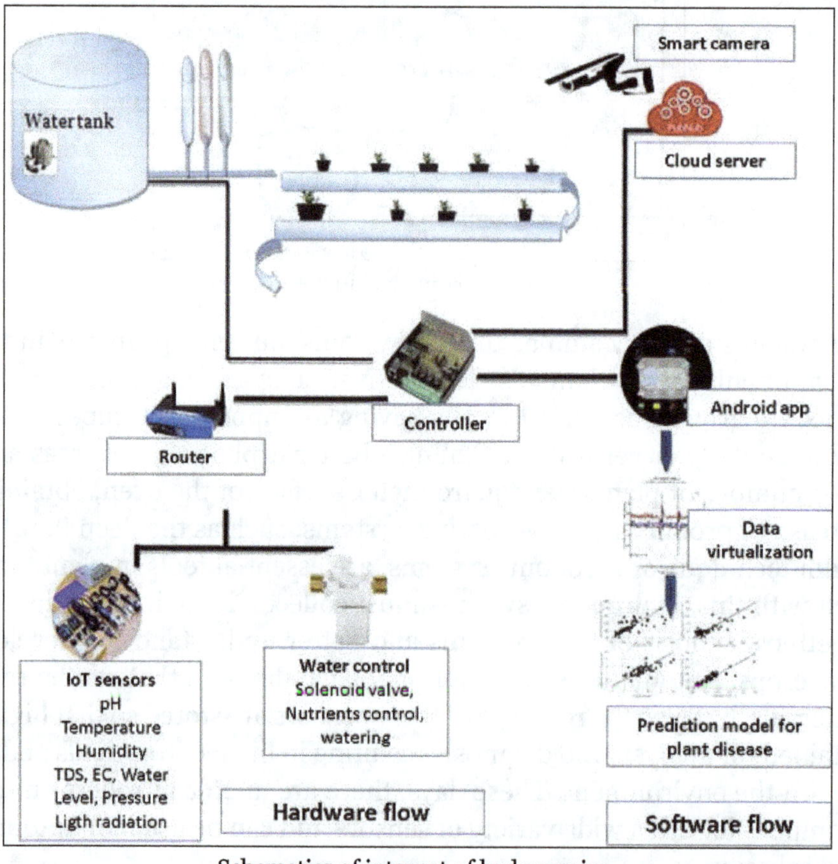

Schematics of internet of hydroponics.

It has been implemented a smart hydroponics system that automates the growing process of the crops using Bayesian network model, which classifies and predicts the optimum value in each actuator to autonomously control the hydroponics farm. Finally, we raise topics related to robotics for hydroponic systems. Hydroponic systems cover approximately 35,000 ha in the world and further research is needed to develop new hydroponic systems to reduce the cost of energy and materials required for crop production.

Automatic grow cabinets for growing plants at home.

Robot for hydroponic systems.

Automation in Hydroponic Systems

All possible variables in root zone must be monitored for automation of the hydroponic system and sensors of pH, the electrical conductivity (EC), light, the ambient temperature, the temperature of the solution, the humidity and the carbon dioxide, the dissolved oxygen and the oxidation–reduction potential must be considered as they directly affect the growth of hydroponically grown plants. The transpiration can be measured with either water ultrasound level sensors or load cells. If the area or volume of culture is large, several sensors must be placed to adequately control the entire crop. Ion sensors (17 essential elements in plant nutrition) are still studied for their durability and stability.

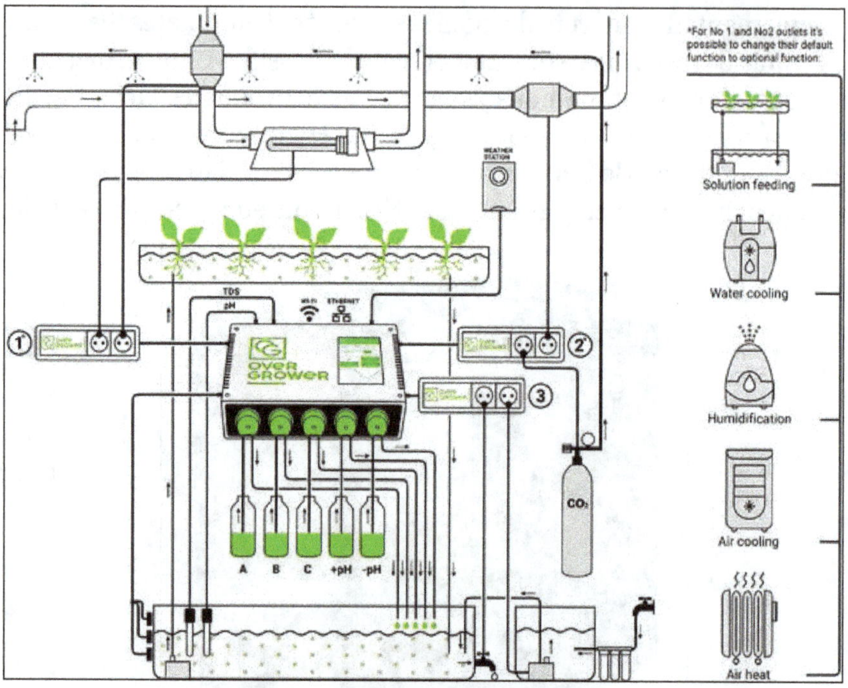

Hydroponics automation system.

Hydroponic System Design with Real Time OS based on Microcontroller

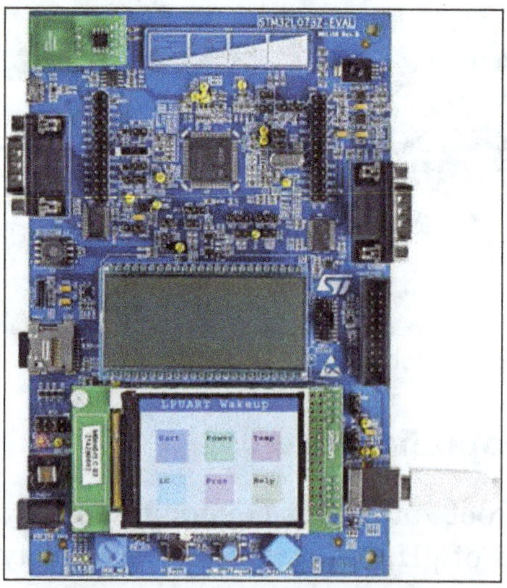

ARM cortex-M4 microcontroller.

It was developed a complete automation hydroponic system for maintaining stable electrical conductivity, pH, growth light and monitoring CO_2, temperature and humidity. The system consisted of an ARM Cortex-M4 microcontroller running ARM embedded operating system, the official real time operating system (RTOS). The system read the pH level and nutrients on the nutrient solution of hydroponics system, as well as the temperature, humidity, CO_2 levels and the light intensity around the system; in addition there were LED light three lines each a different configuration on each line that was used for the lighting of plants and light color was selected and the system data were saved in SD Card.

In addition, the system was capable to control desired concentration level with variation of less than 3%, pH sensor showed good accuracy 5.83% from pH value 3.23–10. Growing light intensity measurement was 105 μmol/m²/s therefore, the lights were turned on at least 17 h/day to fulfill plant light requirement. RTOS gave good performance with latency and jitter less than 15 μs, system overall show good performance and accuracy for automating hydroponic plant in vegetative phase of growth. If the system was turned on, the computer program turned off three pumps (stir pump, water pump and the dosing pump). After initialized an LCD module, then initialized serial and serial to PC for CO_2 sensor. The program read the system configuration data that were stored in the SD card and initialized global variables with the configuration. Later program will update the LCD display. The program started up the sensor to read data sensors, push button to read the buttons provided, mixing to perform compounding nutrients hydroponics, timer flush to set watering plants, timer lights to regulate time lighting plants by LED lights. A timer was started up for minimum water and a sensor to detect the presence of water in nutrients within of a container. Once the water was activated, then timer watered the plants. The pH sensor recorded the initial pH value in the solution, then adding the pH solution up 5 mL and compared the pH sensor measurements obtained with the instrument.

A total of 600 s was taken by DHT22 humidity sensor sampling every 30 s and the readings were compared with the measuring instrument. Twenty-five minutes were taken by MH-Z19 sensor with readings every minute in rooms, results were compared with measuring instruments. The system initially provided nutrients for 5 mL and then the system recorded and calculated the amount of nutrients needed. A distance of 30 cm from LED to plant hole was settled to use a meter for Quantum PAR (photosynthetically active radiation). A LED coefficient was obtained by dividing average light intensity (ALI) with lux. The coefficient of LED and ALI can be used to find daily light integral (DLI) during 17 h. RTOS performance was obtained using square wave input signal and measuring input signal versus output signal delay using oscilloscope. The difference of humidity data retrieval between DHT22 sensors and measuring devices was very small. CO_2 data retrieval between MH-Z19 sensor and measuring devices at room had a difference for each room relatively equal amount, then for the sensor MH-Z19 in this case with a correction factor, so the results obtained are close to the results of measuring instruments. A correction factor of 260 ppm was used for the MH-Z19 sensor against the initial value. The growing light intensity was measured at 25 cm from light source. The result showed each growing light produces difference intensity ranging from 6.03 to 10.74 μmol/m²/s, to fulfill plant light requirement on at least 17 h day^{-1}. From the results of pH meter sensor, the difference obtained is so small so that the pH meter sensor can be used to read the pH suitably. When taking data for RTOS experiment, the main programs still running while the experiment still ongoing.

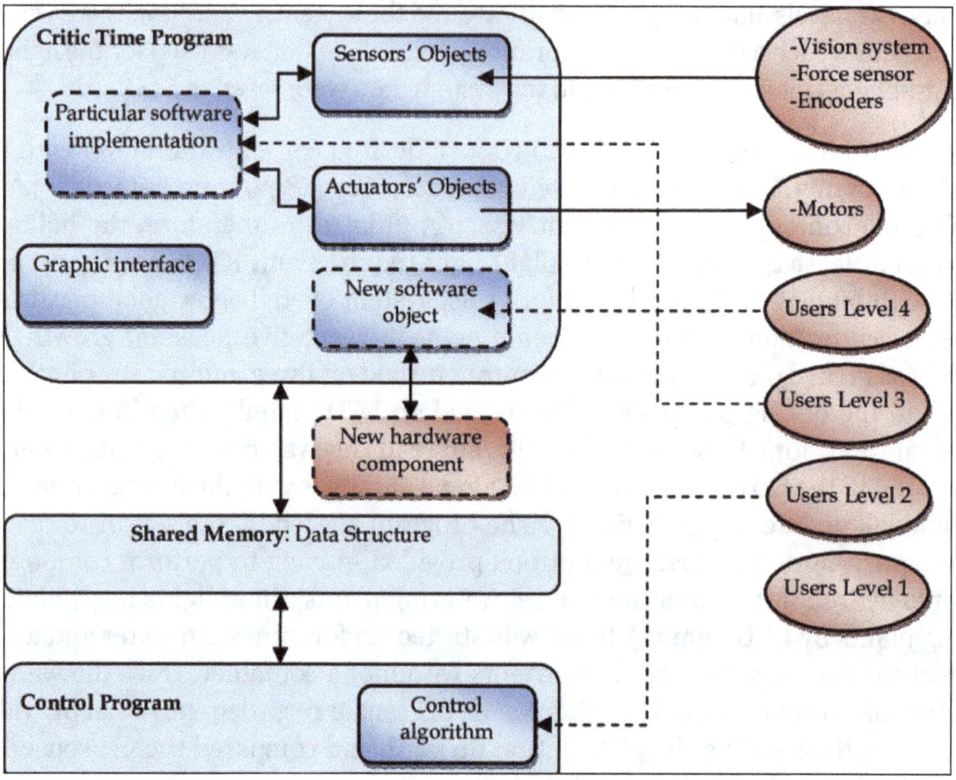

Block diagram of system.

The yellow signal is a given signal and the other signal is a signal output from each thread. Time latency (in microseconds) was very small. The results showed that hydroponic automated system performed well. RTOS ran all the tasks with a latency less than 15 μS. Environment sensor overall showed good result, temperature reading error was less than 4%, humidity reading less than 5.36% and CO_2 sensor accuracy was calibrated 260 ppm from initial value. System was capable to mix nutrients in 80 s with error less than 3.48%. Light intensity measurement showed different result for different color spectrum in order to fulfill daily light plant requirement we need to turn on the light at least 17 h day^{-1}. The vegetable grew well and can be harvested in 5 weeks.

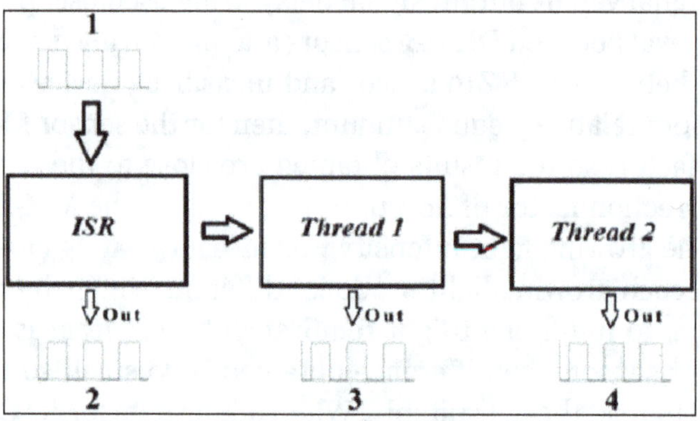

Experiment of RTOS.

pH Fuzzy Logic Control System for Nutrient Solution in Embedded and Flow Hydroponic Culture

The fuzzy-based control system was developed for maintaining a proper acidity level of nutrient solution used in potted flower cultivation of Chrysanthemum embedded and flow hydroponic cultures. Two control valves maintained the nutrient solution pH at a desired set point as follows: (1) acid valve (to manage the addition of acid solution necessary) and (2) base valve (to keep the addition of base solution necessary). The developed control algorithm was based on membership functions of fuzzy arrangement.

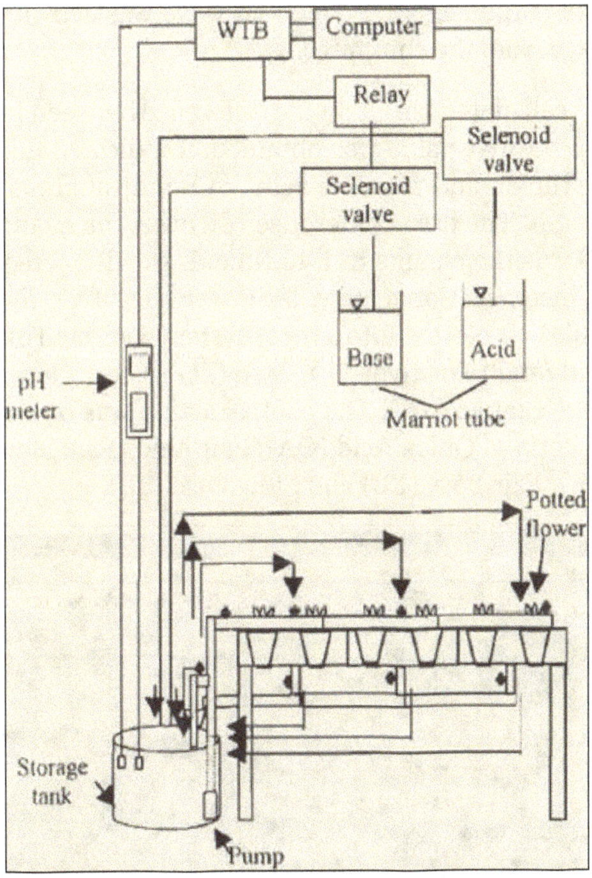

Embedded and flow system with pH control system.

Fuzzy rules had 21 linguistic statements to achieve smoothness, by trials and errors using the membership functions based on the operator skills and experience. The fuzzy logic controlled nutrient solution pH and increased the smoothness of the pH the during control course. The culture vessel consisted of six blocks, each of which containing four potted flowers. The nutrient solution flows into and fills the cultivation bench until a certain level, 5–10 cm from pot base. The embedded system kept the plant growth media in 10 min, before it then flows back into the tank and flows into the next block. The flow rate of the nutrition used in this experiment was 2.4 L min^{-1} and the measuring apparatus was Hanna pH-meter.

The control system maintained 0.3 M H_3PO_4 and 0.4 M KOH, which flowed constantly from Marriott tube. The valve used was of solenoid type with 1/8 in. in diameter. Calibration of the pH-meter was done on voltage basis using PCL-812PG interface. Marriott tube was also used to calibrate the flow rate as well as on the relay circuit. The measurement result of the pH of nutrient solution was in the form of DC voltage and was transferred to 88 shunt circuit in order to get input voltage at a range of 0–5 V conforming to the working voltage of the PCL-812PG interface. This voltage became the reference digital signal for the computer to conduct data processing with control program. The output of the control action was the duration of the solenoid valve opening depended upon the input signal. A solenoid valve was activated by a relay circuit, which obtained voltage from the computer.

Process error (E) was calculated based on the difference between the set point (Sp) and the actual pH. If an E positive value was obtained, it indicated that the position of the actual pH was above the Sp and negative value of E indicated that the position of the actual was under the Sp. The error difference (dE) was the change in E to time. The error difference (dE) was the change in E to time. If the dE were positive, the error E had the tendency to increase. Conversely, if dE were negative, the error E decreased. Every numeric variable was plotted into a fuzzy system consisted of Large Positive (LP), Fair Positive (FP) and Small Positive (SP), Zero (ZO), Large Negative (LN), Fair Negative (FN) and Small Negative (SN). The control action was based on decision matrix in which there are criteria of Quick Acid (QA), Fair Acid (FA), Slow Acid (SA), Neutral (ZO), Quick Base (QB), Fair Base (FB) and Slow Base (SB).

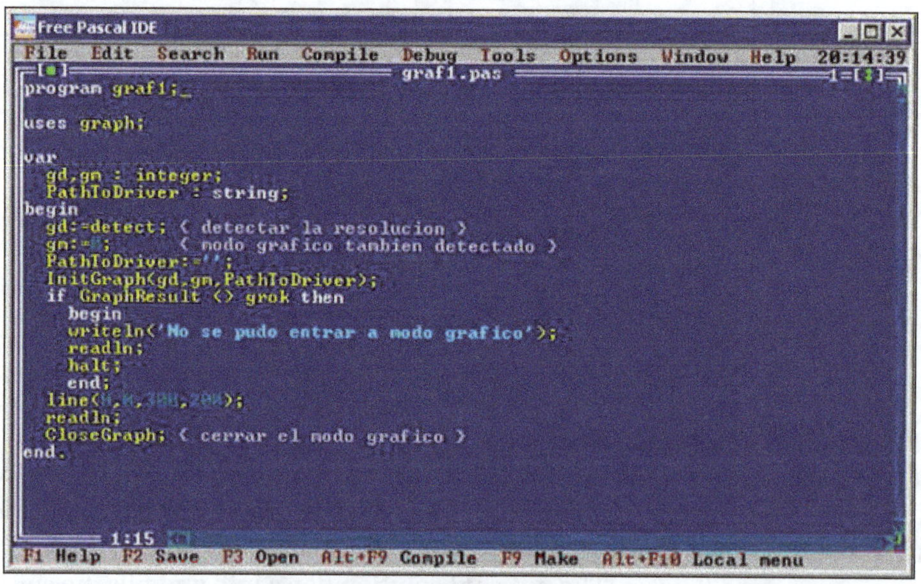

Pascal environment.

The measurement result of the pH of nutrient solution was in the analog form of DC voltage and was transferred to defuzzification by means of weighting to the absolute membership value from every label with the membership degree obtained. A change in valve opening time, either for base tube or acid tube was due to the final output of the fuzzy. The computer program for the control system was developed using the Pascal language in DOS environment. The output voltage from the PCL-812PG had a range of 0–+5 V. The debit of the base and acid flows from the Marriott tube was kept constant at 1.3 eels for base solution and 4.3 $cm^3 s^{-1}$ for acid solution. There were differences in the heads of the air inlet and outlet at the Marriott tubes for base and acid solutions, respectively. The initial pH of the solution was above the set point and kept on moving to reach the pH = 6. To change the pH solution from 7.0 to 6.0, it took 26 s and a 100 s to increase pH from 6.0 to 7.0. That indicated that at the same period of time $[H^+]$ freed by the H_3Po_4 acid was more than that of the ion $[OH^-]$ freed by the KOH base.

The supplied voltage from PCL to relay circuit was on when the voltage reached 1.4 V and off when the voltage decreased to 1.1 V. The pH of the nutrient solution in first block can be controlled to approach the set point of pH = 6. To decrease the pH toward the set point it requires 68 s. After reaching the set point. The pH of the solution did not change very much due to the small change in $[H^+]$ concentration. Moreover, the straight line approaching the set point tendency of the error curve during the control indicates that the fuzzy logic control can maintain the solution pH at the set point. An overshoot not occurred in this pH control. The nutrient solution pH in second and third block can be controlled faster than in first block. The same phenomena occur in third block and the following blocks. In that manner, the set point indicates that the fuzzy logic control can maintain the solution pH at the set point. Both of valves frequently open in turns since the control load was still high at the start up. This frequency decreased at the following blocks.

Hydroponics in Combination with an Automated Drip Irrigation System

A grapevine rootstock in hydroponics in combination with an automated drip irrigation system was developed, which consisted of a hardware and software of the automated hydroponics system for grapevine in pots. Each pot had the same amount of fertilizer and the drip irrigation system was used. It was also constructed a time-based closed loop hydroponics and used a microcontroller for supplying the water to the pots. The irrigation system consisted of a 200 L water storage tank, containing Hoagland solution, which was modified to regulate the optimum 6.2 pH and electrical conductivity levels between 1.0 and 1.3 dS^{-1} for the green cuttings of grapevine. Nutrient solution has been renewed when its EC level reached to 1.5–1.8 dS^{-1}. A submersible pump operating at 12 DCV was installed inside the water storage tank. A steel structure was built to keep pots at height of 1.5 m from the ground level of the greenhouse and water storage tank and controller circuit were installed immediately under the pots. The excess water was easiest to return to the reservoir by the drainage pipes connected to the drainage holes of the pots. Electrical conductivity of the irrigation water (ECw) was measured by an EC59 meter. Pots were irrigated with the same amount of nutrient solution. The required water was supplied by using 16 inches of diameter pipes with 4 $L h^{-1}$-drippers at a spacing of 33 cm, with three drippers serving each pot. Some connection apparatus and valves were used in the irrigation system to integrate all items.

Grapevine experimental setup.

At the beginning of the test, all substrates were filled up to field capacity, then the automated system started irrigation at 4 h intervals and run the submersible pump only 1 min throughout the whole growing season so that this irrigation management kept the soil moisture at the level of field capacity in each substrate since excess water was drained to the reservoir back after each irrigation event. The controller circuit, in which main power supply was 12 DCV, providing power to the controller and relays, but it was reduced to 5 DCV for microcontroller by using a regulator of 7805 and relay.

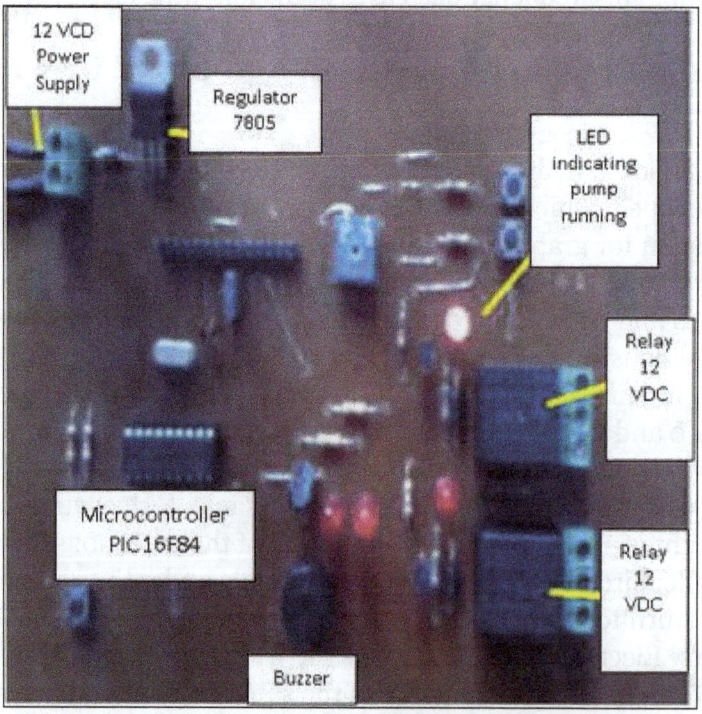

Controller circuit.

The program providing the automation in the hydroponics system was simple and basic and very easy to load into the memory of the microcontroller, which repeated the actions throughout the whole growing season. The dosage of water was determined according to the pumping time of water. The microcontroller switched on relaying to pumping water to the root territory only for 1 min. After that, supplying of water has been stopped to the pump and then waited for 4 h of interval for the next irrigation session. The system took over the irrigation events successfully for the whole growing season. The system conveys a properly balanced nutrient solution to the plant root area. The system saved water and fertilizer, but the water level in the reservoir must be checked with 2- or 3-weeks interval or water level sensor should be added to the controller circuit. Perlite due to its characteristics has more advantages as being used in the hydroponics system as compared with peat and peat + perlite (1:1, v:v). This system can be used for small producers from small hydroponic systems.

Automatic System for Hydroponics

HydroAS could produce fodder in 6 days. The system controlled autonomously the desired agronomic conditions for production and fodder flow. The automatic solution comprised: the mechanical structure, the mechanical and hydraulic components, and the control system to automate the hydroponic automatic system. The mechanical structure consists of the following parts: (a) mechanical structure of six storeys; (b) conveyor to exit the produced fodder off the system; (c) two elevators, at each top of the six storey structure; (d) a fodder sowing system, which placed the seeds in the trays; (e) two pushers at each of the elevator, which pushed the trays in the structure and (f) unloading system, which extracted the finished fodder in the trays.

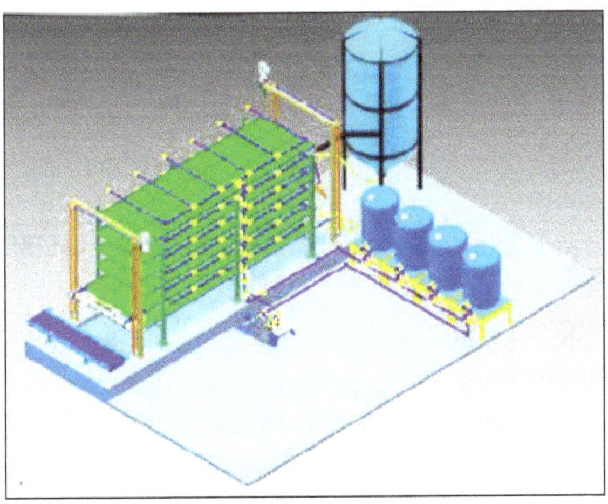

Fodder production system.

The electrical components comprised the power, sensor and actuators circuits. The power circuit consisted of depicted the protections and transformers to obtain 24 DCV to supply the S7–300 programmable logic controller (PLC), sensors and the command circuit, which were mainly digital, inductive and magnetic, that indicate start/end limits for the actuator's movement such as level sensors applied in the nutrients and water reservoirs. The actuators used were (a) two motors for the vertical movement of the elevators; (b) two pneumatic cylinder for the pusher; (c) two worm motors for the sowing platform and (d) two gearmotors for the rotational joint of the sowing and unloading platforms. The hydraulic system supplied the nutrients and the flow of water in the system and consisted of two irrigation pumps to generate redundancy. The hydraulic circuit comprised the valves, nutrients and water reservoirs, the six storeys pipeline with the irrigation micro-jets, and the 2 m^3 water return reservoir with the two redundant pumps. The chosen microspray jets operate at 1 bar and have the capacity of 1 L h^{-1}, with 0.8 m maximum spray diameter area. The pipeline structure consisted of six storeys and watering pumps performed system irrigation three times. The system controls the actuators of the mechanical structure. The nutrient solution control is also performed in the PLC, to control the pH and electric conductivity, while mixing the nutrients and the control sequence of the trays in the hydroponic system.

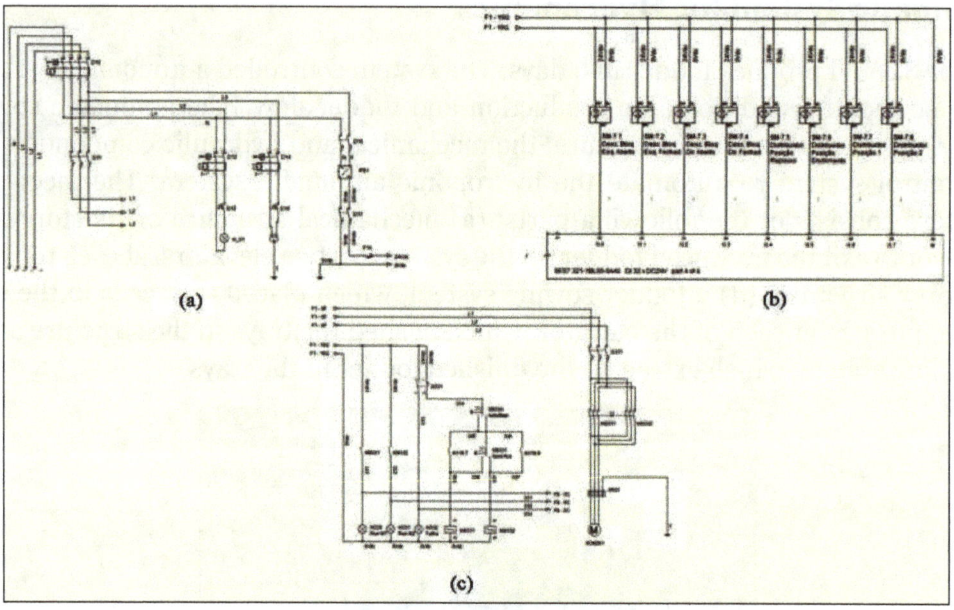

Power circuit (a), sensor circuit for the unloading part (b) and actuator circuit for the elevator one (c).

This sequence definition is a high-level control task, while the low-level actuator control is performed in inner loops and is programmed directly on the motor drives. The system starts when the first storey is filled with trays. When the seeds are placed in the tray, in the first storey near a first elevator, the next step is to elevate the tray and push it to the structure in the second storey. On the other side of the structure, a second elevator, receives a tray that was pushed as consequence of the previous movement. This tray is then elevated to the next storey. This process is repeated until the tray reaches the end. When this happens, the elevator number two descends to the first storey and unloads the produced fodder to the conveyor. After unloading this tray, it is pushed to the first level for washing, and the next 6 days cycle then starts, to produce new trays full of fodder. This system was simulated in Matlab SimMechanics, showing its proper operation for the mechanical and electrical parts. The development phase of the fodder was tested, and validated, which benefits the agricultural holding.

Expert System-based Automation System

HES was developed to minimize the labor force used in the process of hydroponics, the total amount of time spent in agricultural process, human-based errors, as well as, the control of hydroponics greenhouse plant production. All these processes are conducted by a computer unit where the relevant programs are loaded. The system defined the values that belong to the input parameters by using the output parameters that are used by the user. The input parameters prepared the optimum growth environment for the plants. User interface controlled the knowledge base of the expert system and entered data and realize operations. All parameters were taken in to consideration in order to create controlled environment exactly. Knowledge base is continuously in a process of improvement and human experts would add new knowledge to knowledge base or modify the existing knowledge heuristics when new situations occurs. The data base was made up of real conditions that summarize the current situation of the problem and quality-value pairs. By all these output parameters level management, the total level grade can be attained, and this can determine the development period of the plant.

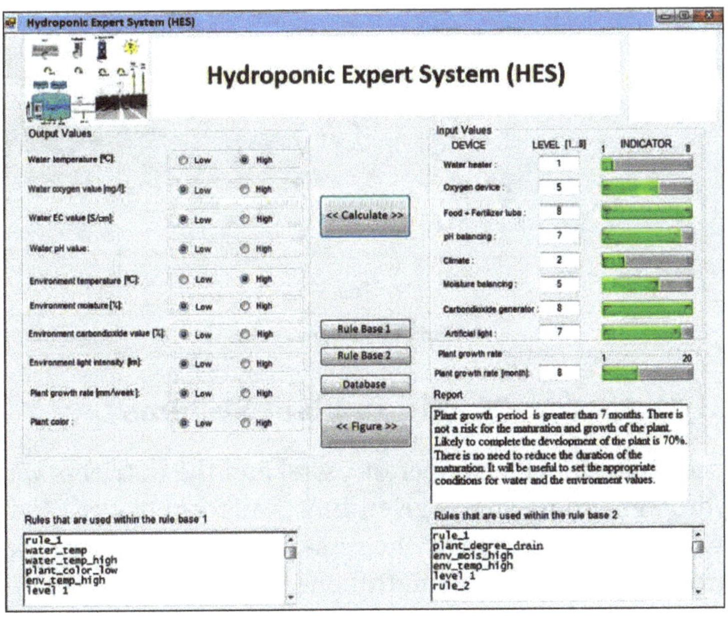

Graphic user interface of HES.

This system had two rule bases. The first one was the rule base that constitutes the input parameters and the second one was the rule base that determines the growth period of the plant. The growth period of the plant is determined by adding the values of plant drain degree, plant nutrition degree, plant deterioration degree, plant photosynthesis degree and plant growth degree. The inference engine, had the function to produce the results that the system needs by using the data in the knowledge base and by interpreting the rules of the system as follows; the user interface of the HES software and the relevant output values taken from the greenhouse system and the sensors are interpreted and translated into linguistic expressions such as low-high. When the system finds a rule that matches the related values in the rule base, it attributes this level value as the level value. It is used for all parameters temperatures, oxygen device, nutrition and the operating levels for water heater, fertilizer tube, pH balance, conditioner, moisture balance, carbon dioxide producer and artificial light are attuned, creating the appropriate conditions for the greenhouse system. Plant growth period is attained from the total of plant drain degree, plant nutrition degree, plant deterioration degree, plant photosynthesis degree and plant growth degree parameters. After all these processes are completed, reports were produced by the system, based on the plant growth period.

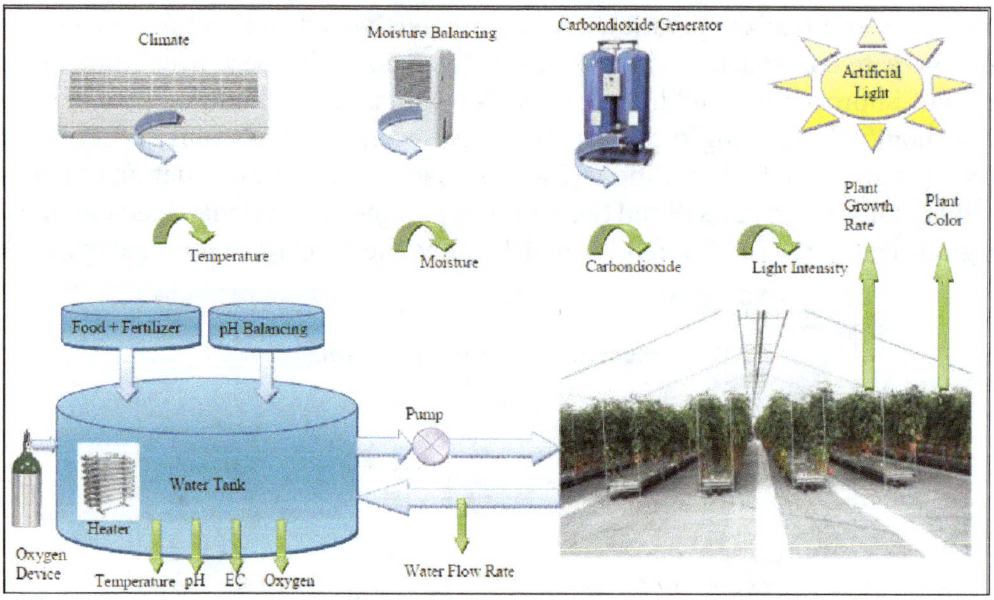

Hydroponic system setup.

Automated Hydroponics Nutrition Plant Systems

AHNPS was placed in a special chamber or vessel and the nutrients were supplied directly to the hydroponic roots at any given time. Microcontroller (Arduino Uno) will control the flow of nutrient solution on the vessel automatically, and the microcontroller can be controlled from Android smartphone. This system had an embedded program module. The microcontroller worked in real time to setup the alarms on nutrient pumps. If alarm is enabled, a relay will be also activated, and then the pump will drain the nutrition solution on the plant. If alarm is deactivated, the relay will be turned off and the pump will stop supplying. Moreover, it has been designed a virtuino application on Android smartphone that serves to check the water level and temperature around the plants. Before starting the design of a virtuino application, it first provides the data storage using the features of thingspeak.com.

Experimental hydroponic system configuration.

The hydroponic flow system starts from the detection of a proximity sensor and a temperature sensor. The sensor will detect the water level in the hydroponic tube and the temperature sensor to detect the room temperature. The detection sensor was connected to a relay that in turn was attached to the microcontroller port. When the relay port pin is lower than the specified height, the water flow will be run on the water pump to irrigate the plant. If the relay of port pin is turned on, it means the water level is above of the specified height, and then the water pump will stop being water, in that manner the water flow was regular. Time and date were displayed at any time of the process in an LCD screen. The pumps are used not only to increase water but also to add nutrients to the hydroponic tube. The water pumps were used for water recirculation and relays used to control both nutrient solution as well drain pumps. The system mechanism worked as follows: HC-SR04 ultrasonic sensor detected the height value of nutrient solution in hydroponic plants by the parameter of the high of water (in cm) unit and the temperature. The LM-35 sensor detected the temperature in Celsius degrees.

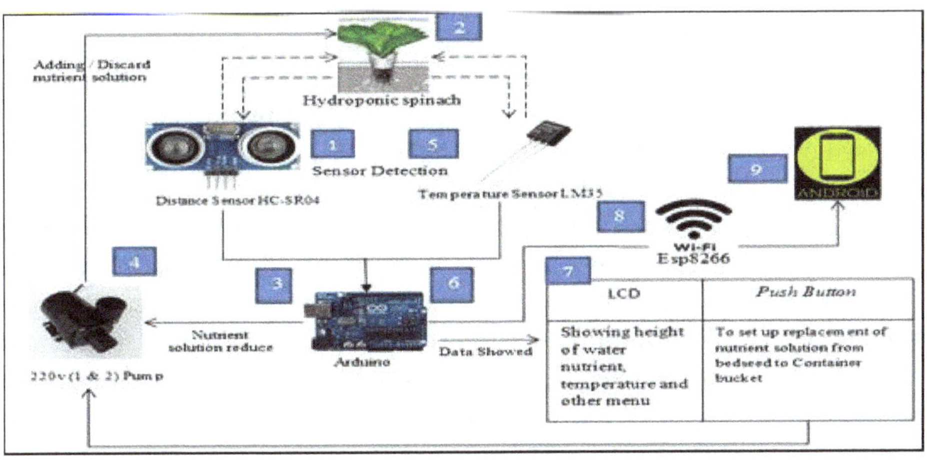

Automated hydroponics nutrition plant system.

The ultrasonic sensors measured the distance of water based on ultrasonic wave. The difference between the transmission time and the reception time became the water distance. The system started working by using of sensors connected to the electrode. The output of this electrode will be forwarded to the microcontroller as an entry point to be processed by the microcontroller. The microcontroller received this voltage signal and compared it with the previous value and decided based on that input signal. Based on this voltage the microcontroller decided whether to drain the water at the pump or not. All the commands on microcontroller (Arduino Uno) can be controlled from a smartphone-based Android. It was observed that this hydroponic plant grows well with proper water and nutrient usage because it is controlled by the microcontroller. The rate of hydroponic plant growth was faster when compared with plants with soil-grown systems. The WiFi module sent the water level information and the temperature values of the plants area. This value was compared to the value in LCD microcontroller and on Android smartphone application. This value is directly obtained from the sensor and sent to the Arduino. After the water level 5 cm in the nutrient tube then the pump stops, and water did not flow anymore. The average temperature for five tests was 28.43°C. The relationship of water height in the nutrient tube with time is recorded continuously by the ultrasonic sensor on several measurements. The maximum level of water was 6 cm in the hydroponic tube. The sensor detected if the water level decrease in hydroponic nutrition tube. If the water level has decreased, then the sensor will perceive what occurs and automatically the water pump will be turned on to increase the water level on the hydroponic nutrient tube.

Hydroponic Management and Monitoring System for an IOT-based NFT Farm using Web Technology

It was created a hydroponic farm management system that could monitor water temperature, water level, higher densities of nutrient solution and the acidity of a nutrient solution using sensors are related and connected to the microcontroller via a website. Hommons used a 20 W solar system, which consisted of a solar cell panel, controllers, battery and DC to AC inverters. The ESP8266 module was used as a communication medium through a wireless network to the internet and integrated with objects that have connection to the internet. Systems can be accessed through the web page using browser based on the server address. The core material of the PVC pipe tool with 3 in. of diameter as his planting medium and ¾ in. of diameter PVC to flow the nutrient solution. The plastic box reservoir served to accommodate any mix of nutrient solution in water. Hommons hardware design relationship of the NFT consisted of sensors, actuator, microcontroller, ESP8266, wi-fi access point, microcomputer (Raspberry Pi) and power supply. In addition, some Raspberry Pi 2 microcomputers served to accommodate the webserver and brokers. Communication technologies on this system using 802.11 or better known as Wi-Fi by using the internet

The power supply using voltage 5 DCV and 2A. Various environmental sensors had been installed to detect any change in the physical or chemical environments and sensors became the input to the process management of NFT. After the user successfully performs the login process so the system redirected the user to the main page heading. There were two buttons on the sidebar navigation. On the main content there were four columns that display data from the sensors-sensors on the NFT hydroponic farming tools, such as nutrient levels, nutrient pH levels, temperature, nutrient and nutrient EC and parts per million (ppm) level. In the navbar a notification function button displayed the alarm or warning to the user while the settings button function settled the system (restart and shutdown) and logout of the system. Automation settings pages were divided into two parts: first part with its own set of pH and ppm values are desirable way entering the value in the textbox. The second automation page contained a selection of plants type which pH and ppm value have been set before, so farmers only need to choose the type of plants that they maintained to grown. After the hardware and sensors on the hydroponic NFT management system were integrated, the sensors (ultrasonic sensors, pH, temperature and EC) needed to be tested to quantify the level of accuracy. The system testing used the original plant samples to find out if the plant is growing well. The plants used in this test are pokchoy, lettuce and kale at the teen age period (after nursery). Plant growth was observed by taking pictures of the plant for a few days.

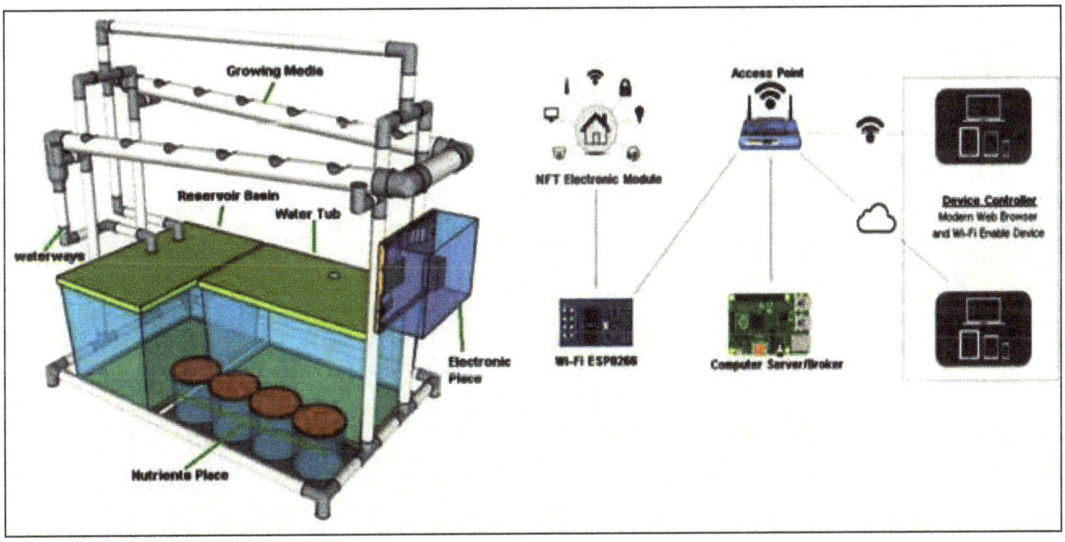

Hommons hydroponic system.

Neural Network-based Fault Detection in Hydroponics

It was developed a fault detection model for hydroponic systems, with a feedforward neural network. Mechanical, sensor and biological faults were considered: a preliminary detection system detected the existence of any faulty situations. Finally, the developed network, only considered two first kinds, mechanical and sensor faults. Biological faults, because of their particularities, were treated separately.

Other model based on a feedforward neural network predicted pH and EC changes in the root zone of Lactuca sativa cv. Vivaldi grown in a deep–trough hydroponic system. The neural net had inputs as follows: pH, EC, nutrient solution temperature, air temperature, relative humidity, light intensity, plant age, amount of added acid and amount of added base and two outputs: pH and EC. A combination of network architecture and training method was one hidden layer with nine hidden nodes, trained with the quasi–Newton backpropagation algorithm which was the most suitable and accurate. The model was capable of predicting pH at the next 20–min time step within 0.01 pH units and EC within 5 μS cm^{-1}. Simpler prediction methods, such as linear extrapolation and the lazy man prediction, value of the previous time step, gave comparable accuracy much of the time, though, they performed poorly in situations where the control actions of the system had been activated and resulted rapid changes in the predicted parameters. In those cases, the neural network model did not encounter any difficulties predicting the rapid changes. Thus, the developed model successfully identified dynamic processes in the root zone of the hydroponic system and accurately predicted one–step–ahead values of pH and EC.

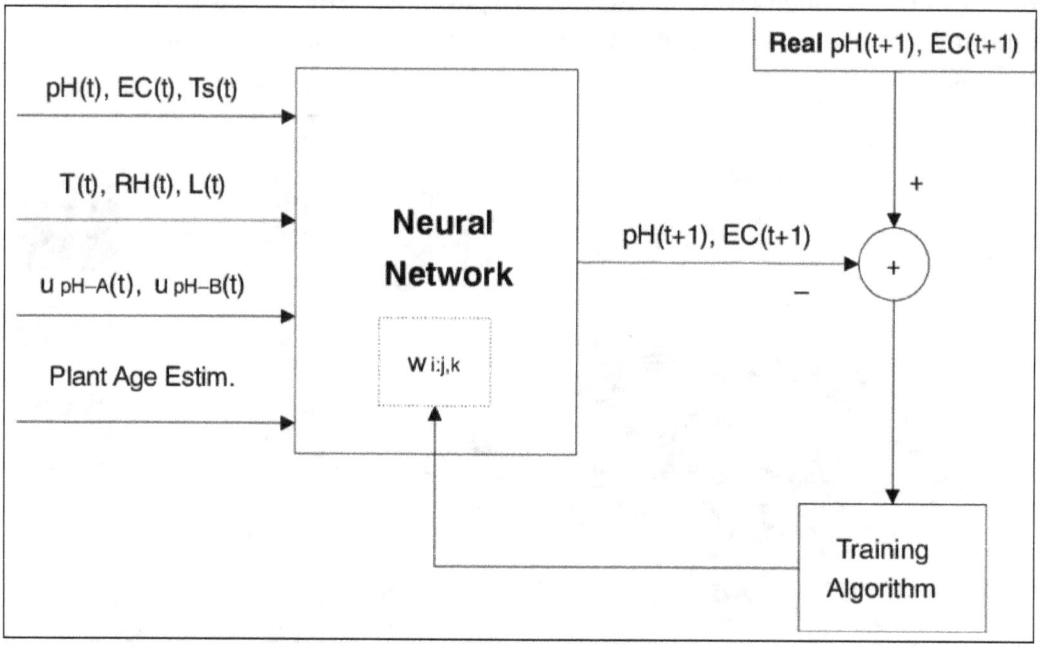

Neural network model inputs and outputs and training process.

IoT-based Intelligent Hydroponic System

An IoT-based intelligent hydroponic plant factory solution called PlantTalk was developed. PlantTalk intelligence could be built through an arbitrary smartphone. PlantTalk was flexibly to configure the connections of various plant sensors and actuators through a smartphone. Python programs for plant care intelligence through the smart phone were convenient. Automatic LED lighting, water spray, water pump and so on were included in the developed plant-care intelligence included and so on. For instance, it was showed that the PlantTalk intelligence effectively lowers the CO_2 concentration, and the reduction speed is 53% faster than a traditional plant system. AgriTalk for a plant factory is an extension of PlantTalk.

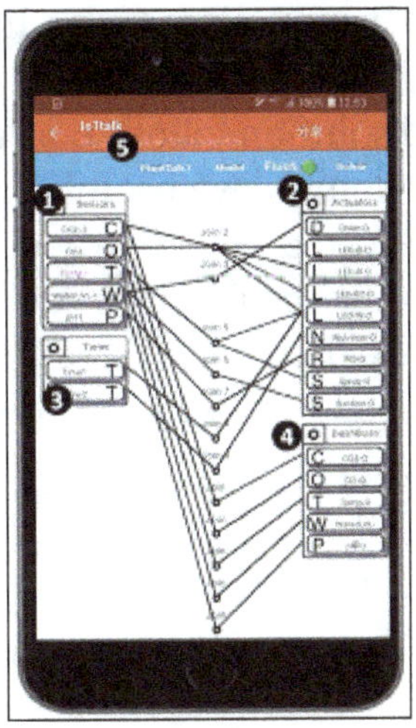

PlantTalk in a smart phone.

Other Technologies used in Hydroponic Systems

It was applied ultrasound and dissolved oxygen supersaturation as external for controlling the growth rate of plants in hydroponics as well as maintaining the product quality. In the case of the leaf lettuce growth in hydroponics with exposure to 28-kHz ultrasound and dissolved oxygen supersaturation up to 36 mg L^{-1} at 20°C and peak-to-peak pressure at 20 kPa or larger worked as the growth inhibitor of the leaves and the roots; in addition, oxygen supersaturation became a growth promoter, without any degradation of chlorophyll in the leaves.

On the other hand, liquid separated reactor and a high voltage power supply based on a 20 kHz inverter neon-transformer were developed to archive the treatment with high energy efficiency, a low initial cost and a low running cost. The performance of the system on bacteria inactivation in the nutrient solution was evaluated in a continuous treatment system operation and the results showed that the standard plate count for background microflora and R. solanacearum is drastically reduced by the plasma treatment and is not detected after 8 days treatment. The nutrient solution was decontaminated by 4 log cycle with plasma treatment under the continuous operation condition.

Other researchers applied electro-degradation (ED) to the culture solution in order to degrade their root exudates and improve growth, yield and quality of strawberry. They used four types of nutrient viz. renewed, non-renewed, non-renewed with direct current electrodegradation (DC-ED) and non-renewed with alternative current electro-degradation (AC-ED). Fresh 25% standard Enshi nutrient solution were changed every 3 weeks interval, with in renewed treatment, while DC- and AC-ED treatment were applied in non-renewed solutions. Significantly greater fruit yield (225.9 g $plant^{-1}$) was obtained from renewed nutrient solution, which was statistically similar to fruit yield in non-renewed solution with AC-ED application. Compared to renewed solution, fruit yield was decreased to about half (114.0 g $plant^{-1}$) in non-renewed solution while non-renewed with DC-ED produced intermediate yield between non-renewed and renewed solution or non-renewed with AC-ED. It was concluded that growth performance was greater in renewed solution followed by non-renewed with AC-ED, while it was decreased significantly in nonrenewed solution with DC-ED similar to non-renewed solution. It was also observed a similar trend in vitamin C content while brix and citric acidity was not varied. Calcium and iron concentration in the culture solution were significantly decreased in DC-ED, consequently their contents were also found lower in crowns and roots compared to other solutions used. The strawberry yield and quality can be improved through application of AC-ED in non-renewed solution.

Robotics in Hydroponic Systems

Robot with Position-based Visual Feedback

RPBVF was developed to act and observe the crop in NFT hydroponic systems. The focus is on the implementation of a position-based visual feedback (PBVF) algorithm in combination with a Microsoft Kinect. AmHydro 612 NFT production unit was 1.8 m × 3.65 m × 0.9 m production unit that stored 144 plants and 144 seedlings and used a closed loop water system. Above the NFT system were placed artificial lights to improve the lettuce growth. The gullies laid on an inclined table, which angle was θ, so that water flows passively to the end of the gullies. Water was collected at the end of the gullies and directed to the water reservoir, where a water pump propelled water to the top of the gullies. To manipulate the plants, the robot was designed as a gantry with four v-groved wheels running on two inverted angle iron tracks (x-axis). On top of the gantry was a carriage that can move back and forth over the gantry (y-axis), this was perpendicular to the x-axis. On the carriage was a mechanism to move an arm up and down (z-axis), down being the negative direction. At the end of the arm was placed a two degrees of freedom gripper which opened, closed and rotated around the y-axis.

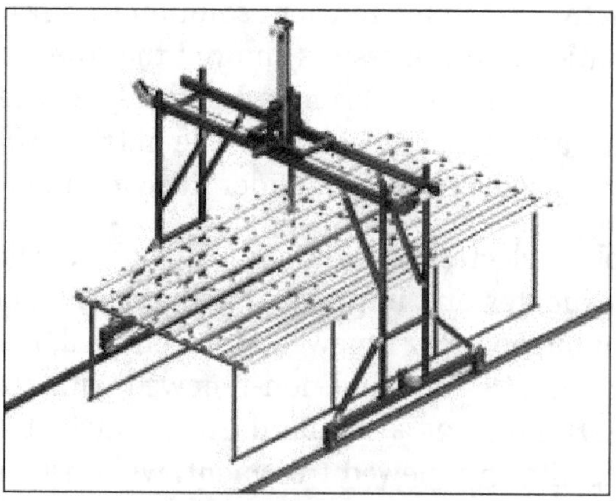

Robot manipulator arm.

The structure is made primarily from aluminum that allows the robot to be adjusted to accommodate different sizes of NFT hydroponic systems. The x-axis was driven by a stepper motor and a chain. A timing belt transmitted the power from the carriage on the gantry to the stepper motor. The arm on the carriage was balanced by a counterweight and was driven by a stepper motor and a chain. Two linear actuators were used to open and close the gripper and the other linear actuator was used to rotate the gripper around the y-axis. All three linear actuators were driven by a 12 DCV relay board that communicates with a Phidgets interface board, which was connected to the main computer, which was running Ubuntu Server 11.04 × 64. The Kinect vision system was mounted on the carriage so that the optical axis was along the negative z-axis. All software was programmed in C++. Every hardware component communicated with its own ROS (Robotic Operating System) node. The main hardware nodes were stepper motor node, gripper node, interface board node, position node and Kinect node. The position node keeps track of the x, y and z-position of the robot and a graphic user interface was designed to provide low level control of the system. A Microsoft Kinect camera was added to the system, which produced two kinds of images, a 640 × 480-pixel RGB color image and a 640 × 480 pixel 11-bit (0–2047) gray scale depth image that was provided by an Infra-Red (IR) sensor. The extraction of the plants required combining classical 2D image analysis techniques and IR-based depth measurement the 3D position. The computer language used was C++ using the Open Computer Vision library. The Kinect was placed on the carriage and is facing downwards (negative z-axis) to ensure the plants in its field of view are at a maximum distance of 1.5 m, because the accuracy of the Kinect decreases quadratically with distance. Up to 1.5 m the accuracy was 10 mm and the precision of the Kinect was 1 mm. The field of view was of 0.8 m × 1.15 m in x and y-direction. It was used a RPBVF algorithm was used to detect plants on the hydroponic system and placed the robot to manipulate plants.

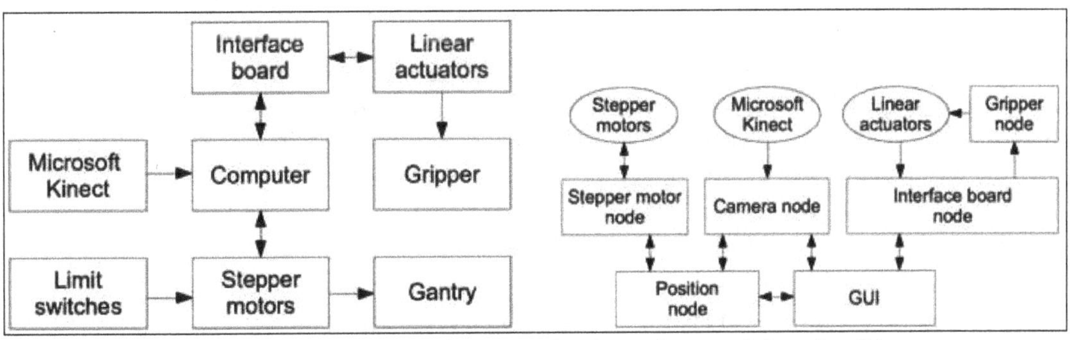

Hardware layout of the robot (a) and software layout of the robot (b).

In the algorithm, first were detected the gullies, because the plants are only located on them. All gullies are oriented along the x-axis and are straight. A probabilistic Hough Transform was used for straight line detection. By filtering the detected lines, the edges of the gullies were identified. After the identification of the edges, the lines were grouped, resulting in a segmentation of the gullies. After filtering, the coordinates of the plants in the image frame were known. Point was defined as the top left corner of the image. The depth information was extracted from the depth image by getting the value at point. The OpenNI driver transforms the IR sensor values into distances in meters by using a fitting function. To reduce the noise, multiple consecutive frames were averaged to calculate the plant coordinates. The plant coordinates form the control input for the robot. The output only depends on the current state and the control input. The open loop control algorithm was used. To be able to pick up a plant, the image frame coordinates had to be transformed to gantry coordinates. To transform the image frame coordinates to gantry coordinates a Garstka and Peters modified transformation was used. Because the Kinect is not located on the gripper, all coordinates have to be offset. These offsets are dependent on the position of the Kinect relative to the gripper. The z-coordinate has to be offset by an extra value, because the NFT table is under an angle of 2.2°. In this transformation, the point is the principal point in pixels of the depth sensor and the focal lengths in pixels were calculated.

The values were quantified by calibrating the Kinect. The position bias was removed by a linear scaling of the x and y-coordinates. To evaluate the performance of the positioning and control algorithm, the x, y and z-position error between the final position of the gripper and the plant coordinate was measured. The final position of the gripper was defined as 20 mm above the center of the cup. On the top of the cup a cross-hair is drawn to mark the center. The initial position of the gripper is defined as the middle between the points of the gripper so the x, y and z-position error can be measured. Each image was analyzed to detect the plants. With these coordinates the robot is heading to the plant. The position error was measured with a ruler at the final stopping position and robot then returned to the same starting position. The gripper must be ±15 mm in x-direction, ±20 mm in y-direction and ±10 mm in z-direction from the center of the cup to allow the robot to pick up the plant. From the images with detected plants the gantry coordinates of the plants were calculated and inputted for the positioning algorithm so that the robot can be positioned to pick up the plants. There were 25 samples evaluated. The performance of the system is within the requirements and the plants could be manipulated on an NFT system.

Growing Mediums in Hydroponics

In hydroponics, the growing medium takes the place of the dirt/soil. Not to provide nutrients, but so the roots can support the plants weight and hold it upright. Just about any inert material can be used as a growing media. Inert meaning that it can't/won't decay or break down quickly, thus providing nutrients to the plants. Hydroponic growing media is simply a soil-less material that is generally porous so it can hold the moisture and oxygen that the root system requires to grow. Non porous materials can be used as well, but watering cycles would need to be more frequent so the roots don't dry out between watering's.

The growing medium won't be able to grow anything on its own. If you placed plants in hydroponic growing media, and watered it with plain water, the plants would starve from nutrient deficiency. It's simply there to help support the plants weight as well as the moisture and oxygen the roots need. The nutrients the plants needs are provided by the nutrient solution, and is what the growing media is watered and moistened with.

Some of the most widely used growing media's include Rockwool, Lightweight Expanded Clay Aggregate (called, Hydrocorn or Grow Rock), Coconut Fiber/Coconut chips, and Perlite or Vermiculite. While there are a lot of materials that can be used as growing media in hydroponics, they can all have very different properties than another type of media. We have even seen the use of hay bales as growing medium to grow tomatoes, using a drip lines on top to drip the nutrient solution onto the hay bales and tomato plant roots.

There isn't one growing media that is better than the rest. Especially with so many different hydroponic system designs possible. However many growers eventually favor one type over others. There are a lot of things to consider when choosing what to use as a growing media. The type of system your growing in, and how you design and build that system is the biggest factor.

While there is no one best growing media for all situations, some growing media's work better than others in different systems. With any hydroponic system, and/or any type of growing media, the goal is still the same. You just need the roots to be moist, not soggy and saturated. If the growing media is saturated and soggy, the roots will suffocate from lack of oxygen. That situation can easily lead to roots dying, and root rot.

Things to Consider about Growing Media for each Type of Hydroponic System

Drip Systems

Drip systems are fairly easy to control moisture in. As long as you design it so it has good drainage, and limit water pooling at the bottom you should be able to keep your growing media from being too soggy. We like to use river rock at the bottom to aid drainage, and keep the growing media from sitting in a pool of water at the bottom.

NFT Systems

NFT systems use a very shallow, but continuous stream of water at the bottom of a channel where the roots wick up moisture. Most NFT systems either use small starter cubes or small 1 inch baskets, then let the roots just hang down into the flowing water. If these cubes or baskets are to close to the water supply, and your growing media can become saturated easily, that combination can lead to "stem rot" if the growing media around the stem is always saturated.

Ebb-flow Systems

Flood and drain systems can vary quite a bit in design. But generally you would want to stay away from any growing media that floats like Perlite and Vermiculite. Each time the system cycles on for the flooding stage the growing media will become weightless, then your plants will loose all their support and want to tip over. Depending on your system you could lower the flood level so floating is minimal, as long as the root-ball can still get plenty of moisture and you don't loose much plant support. Plant support may not be as much of a factor if growing vine type plants like tomato's, peas or melons that would be tied to a trellis as well.

Also with flood and drain systems, and depending on the type of growing media you choose, you want to make sure you have good drainage so that the growing media isn't continually saturated. Grow rocks won't wick up much moisture, but we like using coco chips a lot because they are inexpensive, but coco chips wick up water if they are sitting in it. So a layer of river rock at the bottom keeps it from sitting in the water.

Water Culture Systems

Water culture systems don't generally use much if any growing media because it is designed so the plants roots are submerged into the nutrient solution itself. So plants are generally started using small starter cubes or small baskets. Starter cubes typically are suspended above the water line, while baskets can suspended either just above or just below the water line. The growing media you choose and how much it absorbs moisture will make a difference. You don't want it to become saturated, you just want it to be moist at the bottom, the top should be dry. The roots will grow downward into the nutrient solution.

Now you may be wondering if it's so bad if the growing media becomes saturated, why won't the roots become suffocated if their underwater all the time in a water culture system. First not all plants do well in water culture systems. Second and most important, a water culture system uses an air pump to generate a lot of air bubbles to the roots under water. The plants get oxygen from these air bubbles directly, as well as the air bubbles increase the dissolved oxygen levels in the water itself.

Aeroponic Systems

Aeroponic systems typically don't use much growing media at all. Aeroponic systems are designed to allow the roots hang in the air while getting frequently getting misted with nutrient solution so the roots don't dry out. Seeds are started in either small starter cubes small baskets, then when their big enough their planed in the aeroponic system. You'll want to make sure the cubes or growing media in the baskets don't become saturated. While almost all the roots will be hanging in air with no chance of becoming suffocated, wet growing media around the plants steam can lead to "stem rot".

Wick Systems

Wick systems are the least used type of system, but with no moving parts, motors or pumps, they rely on wicking up moisture into the growing media and to the plants roots through a piece of fabric. With wick systems you'll want to use a growing media that absorbs and holds onto moisture easily. You can control the amount of water getting to the plant by using a larger/wider wick, or more than one.

List of Different Types of Growing Media for Hydroponics

Rockwool

Rockwool is one of the most common growing media's used in hydroponics. Rockwool is a sterile, porous, non-degradable medium that is composed primarily of granite and/or limestone which is super-heated and melted, then spun into small threads like cotton candy. The rockwool is then formed into blocks, sheets, cubes, slabs, or flocking. Rockwool sucks up water easily so you'll want to be careful not to let it become saturated, or it could suffocate your plants roots, as well as lead to stem rot and root rot. Rockwool should be pH balanced before use. That's done by soaking it in pH balanced water before use.

Grow Rock (Hydrocorn)

Grow rock is a Lightweight Expanded Clay Aggregate (L.E.C.A)., That is a type of clay which is super-fired to create a porous texture. It's heavy enough to provide secure support for your plant's, but still light weight. Grow rocks are a non-degradable, sterile growing medium that holds moisture, has a neutral pH, and also will wick up nutrient solution to the root systems of your plants. Hydrocorn grow media is reusable; it can be cleaned, sterilized, then reused again. Although on a large scale, cleaning and sterilizing large amounts of grow rocks can be quite time consuming. Grow rock is one of the most popular growing medium used for hydroponics, and just about every store selling hydroponics supplies carries it.

Coco Fiber

Coco Chips

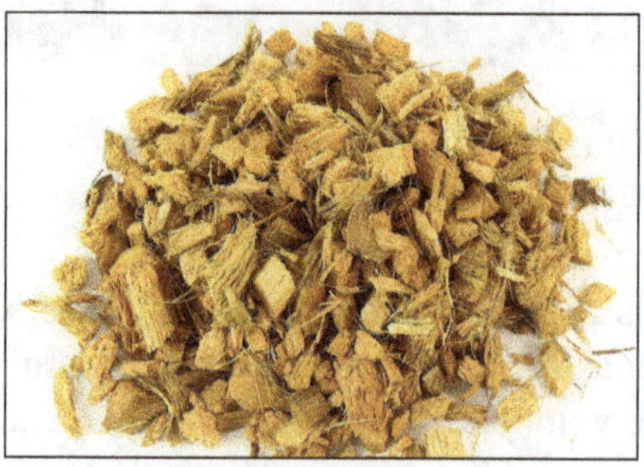

"Coco coir" (Coconut fiber) is from the outer husk of coconuts. What was once considered a waste product is one of the best growing mediums available. Although coco coir is an organic plant material, it breaks down and decomposes very slowly, so it won't provide any nutrients to the plants growing in it, making it perfect for hydroponics. Coco coir is also pH neutral, holds moisture very well, yet still allows for good aeration for the roots. Coco fiber comes in two forms, coco coir (fiber), and coco chips. They're both made of coconut husks, the only difference is the particle size. The coco fiber particle size is about the same as potting soil, while the coco chips particle size is more like small wood chips.

The larger size of the coco chips allows for bigger air pockets between particles, thus allowing even better aeration for the roots. Also if your using baskets to grow your plants in, the chips are to big to fall through the slats in the baskets. Both the fiber and chips come in compressed bricks, and once soaked in water it expands to about 6 times the original size. Coco fiber does tend to color the water, but that diminishes over time. And you can leach out most of the color if you soak it in warm/hot water a few times before use.

Perlite

Perlite is mainly composed of minerals that are subjected to very high heat, which then expand it like popcorn so it becomes very light weight, porous and absorbent. Perlite has a neutral pH, excellent wicking action, and is very porous. Perlite can be used by itself, or mixed with other types of growing media's. However because perlite is so light that it floats, depending on how you designed your hydroponic system, perlite by itself may not be the best choice of growing media for flood and drain systems.

Perlite is widely used in potting soils, and any nursery should carry bags of it. However perlite is sometimes also used as an additive added to cement. You may find it for a better price with the building supply's, and/or at places that sell concrete mixes and mixing supply's. When working with perlite be careful not to get any of the dust in your eyes. Rince it off to wash out the dust, and wet it down before working with it to keep the dust from going airborn.

Vermiculite

Vermiculite is a silicate mineral that like perlite, expands when exposed to very high heat. As a growing media, vermiculite is quite similar to perlite except that it has a relatively high cation-exchange capacity, meaning it can hold nutrients for later use. Also like the perlite, vermiculite is very light and tends to float. There are different uses and types of vermiculite, so you'll want to be sure what you get is intended for horticulture use. The easiest way to be sure is to get it from a nursery.

Oasis Cubes

Oasis Cubes are similar to Rockwool cubes, and have similar properties. But oasis cubes are more like the rigid green or white floral foam used by forests to hold the stems in their flower displays. Oasis cubes are an open cell material which means that the cells can absorb water and air. The open cells wick moisture throughout the material, and the roots can easily grow and expand through the open cell structure. While oasis cubes are usually used as starter cubes for hydroponically grown plants, they also have bags you can fill your growing containers with. While oasis cubes are similar to rockwool, Oasis cubes don't become waterlogged as easily as rockwool cubes. Even so don't let it stay in constant contact with the water supply, or you'll still have water logging issues.

Floral Foam

Floral foam can be used as a growing media in hydroponics as well, and is similar to the oasis cubes, though the cell size is larger in the floral foam. Depending on the type of hydroponic system you're using, and how you designed it, you may notice a couple of problems with using floral foam. First it can crumble easily and that can leave particles in your water. Second you'll want to be sure it doesn't get water logged. Floral foam absorbs water easily, so make sure it isn't in constant contact with the water supply.

Growstone Hydroponic Substrate

Growstones are made from recycled glass. They are similar to grow rocks (hydrocorn) but are made of clay and shaped marbles. Growstones are light weight, unevenly shaped, porous, and reusable, they provide good aeration and moisture to the root zone. They have good wicking ability and can wick water up to 4 inches above the water line. So you'll want to make sure it has good drainage or is deep enough so it doesn't wick water all the way to the top. Otherwise like with the growing media in any hydroponic system, if the top of the growing media is continually wet, you may have problems with stem rot. While they are made from recycled glass, they're not sharp and you won't get cut from it, even if they break.

River Rock

River rock is common and easy to find in home improvement stores, as well as even pet supply's stores (with the fish and aquariums). River rock is fairly inexpensive (depending on where you get it from), and comes in many different sizes. River rock is rounded with smooth edges from tumbling down the river. Though manufactured river rock is rounded using large mechanical tumblers, it has the same end result with smooth edges.

You can use regular rocks from your back yard in hydroponic systems as well if you don't mind the jagged edges. Just make sure to clean and sanitize them before using it.

Just spray all the dirt off of the rock using the jet spray from your hose to clean it, then soak it overnight in bleach water to sanitize it. Then just rinse and use. Though using rock as a growing media is inexpensive and easy, it will get heavy quickly, so you won't want to move it later.

River rocks are not porous, therefor it doesn't hold and retain moisture in the root zone of hydroponic systems. Rock is uneven so it has a lot of air pockets between the rocks so the roots can get plenty of oxygen, but water easily drains down to the bottom. Rock won't wick up moisture either, so you will need to adjust your watering schedules so the roots don't dry out between watering's. You can mix in some coco chips or other growing media that holds moisture with your rock to aid it in holding onto moisture longer.

Because of the good drainage properties of rock, it's very good to use to aid in the drainage of other hydroponic growing media's that might otherwise become saturated from sitting in water. A layer of rock at the bottom of the growing container will keep your growing media from sitting in water at the bottom of the container, keeping it from being saturated.

Pine Shavings

Pine shavings are an inexpensive hydroponic growing media as well, and a lot of commercial growers use it. Generally for large scale hydroponic drip irrigation systems. Don't confuse pine shavings with saw dust. Saw dust will become compact and water logged easily. You'll want to make sure your pine shavings were made from kiln dried wood, and does "NOT" contain any chemical fungicides. Kiln dried to burn off all the sap in the wood that is bad for the plants. Most pine shaving products would be kiln dried to begin with.

Good source to find pine shavings are pet supply stores. Its used for things like hamster and rabbit bedding. Just make sure to read the package to be sure it doesn't have any chemical additives like fungicides or odor inhibitors. You should be fine if it states it's organic. Another good cheep source for pine shavings is at feed stores, it's also used as bedding in horse stalls and they sell it by cubic yard. If you have a choice get the largest partial size you can. The larger the air pockets between the shavings, the better aeration to your roots.

Pine shavings are a wood product, so they absorb water easily, thus can become water logged easily. So make sure you have good drainage so the shavings don't sit in water. If there is a possibility of it sitting in water, a layer of rocks at the bottom will aid drainage greatly.

Composted and Aged Pine Bark

Pine bark is one of the first growing media's used in hydroponics. It was generally considered a waste product, but has found uses as a ground mulch, as well as substrate for hydroponicaly grown crops. Pine bark is considered better than other types of tree bark because it resists decomposition better, and has less organic acids that can leach into the nutrient solution than others. Bark is generally referred to as either fresh, composted, or aged.

Fresh bark uses up more nitrogen as it begins to decompose, so commercial growers generally compensate by adding extra nitrogen to the nutrient solution. During the composting process, nitrogen is added to the bark, and mixing it in while breaks down. So nitrogen issues are far less of a concern with composted pine bark. Ageing is a similar process, but has less nitrogen added to it, so it's better than using fresh bark, but not as easy as the composted bark. Pine bark can be found at places that sell ground mulch, as well as ground mulch for playgrounds.

Poly (Polyurethane) Foam Insulation

Polly foam is not commonly used in hydroponics, and hydroponics stores don't carry it. But has been used as an alternative to using rockwool or oasis cubes as starter cubes with great results. Polly foam is cheap and easy to find. Any hobby store or place that sells fabrics should carry it. It's most commonly used as furniture foam, and is also referred to as "foam batting". It comes in sheets or rolls of different sizes and thickness. You can make your own starter cubes for about one penny each using the poly foam if you get the one or two inch thick sheets/rolls and cut them into cubes.

Water Absorbing Crystals and Polymers

Water absorbing polymer crystals have been around for quite a while, and are used in many industry's. Everything from baby diapers, to the sports industry where they are used in cloth rags they can where on the head or neck to keep cool. They are also used in gardening where the crystals are mixed into the soil to help retain moisture in the soil. Florists use them in vases to keep flowers fresh, and the colored ones make for a nice decorated display.

The crystals expand to many times their size as they soak up water. One pound of the crystals can hold as much as 50 gallons of water. The crystals come in many sizes, everything from a powder, to marble and even golf ball size. Depending on the size of the crystals they can take more than an hour or two to fully absorb. When they are full of water they look and feel like a glob of jello. Once they dry out, they can be stored and reused again over and over.

The water absorbing polymer crystals are not a common hydroponic growing media, but like everything else, it's growing in popularity. Mostly due to their increased availability. They are quite inexpensive, and reusable. However used alone by themselves they don't allow the roots to get much oxygen/air. Being like jello they pack together and fill the air pockets. The larger size crystals are better suited for use in hydroponics. The larger size helps retain some of the air pockets between the crystals. Also by mixing some river rock or other similar growing media with the crystals will help increase the air pockets between the crystals.

Using the polymer crystals for hydroponics allows for some of the simplest hydroponic system designs. Even on the slimmest of budgets. Simply soaking some water absorbing crystals in nutrient solution, then setting them in a container and placing your seedlings in it, you've got a hydroponically grown plant. You don't need any pumps.

Just make sure there are holes in the bottom of your container, and just place your container in nutrient solution once or twice a week to re-hydrate the crystals.

You won't find water absorbing polymer crystals in hydroponic shops, but they are easy to find. Because of their popularity, most large nursery's carry them as soil amendments. Also if you do a search for them online, you'll fine hundreds of places selling them.

Sand

Sand Is actually a very common growing media used in hydroponics. It's the main growing media used at the Epcot Center Hydroponic Greenhouse in Florida. Mainly for their large hydroponically grown plants and trees. Sand is like rock, just smaller in size. Because the particle size is smaller than regular rock, moisture doesn't drain out as fast. Sand is also commonly mixed with Vermiculite, Perlite, and or coco coir. All help retain moisture as well as help aerate the mix for the roots.

When using sand as a growing media you will want to use the largest grain size you can get. That will help increase aeration to the roots by increasing the size of the air pockets between the grains of sand. Mixing Vermiculite, Perlite, and or coco coir with the sand will also help aerate. You will also want to rinse the sand well before use to get as much of the dust particles out of it as you can. One big downside to using sand as a growing media for hydroponics is that it is very heavy. 3-4 gallons of wet sand can weigh up to 50lbs. So you won't want to be moving it once you get it set up. Or use it in a ratio of something like 20%-30% sand and the rest Vermiculite, Perlite, or another type of growing media to reduce weight.

Rice Hulls

Depending on your location, rice hulls may be readily available. It's a byproduct of the rice industry. Even though they are an organic plant material, they break down very slowly like coco coir, making them suitable as a growing media for hydroponics. Rice hulls are referred to as either fresh, aged, composted and parboiled, or carbonized. Fresh rice hulls are typically avoided as a hydroponic growing media because of the high probability of contaminants such as rice, fungal spores, bacteria, decaying bugs, and weed seeds. Parboiled rice hulls (PRH) is done by stemming and drying the rice hulls after the rice has been milled from them. This kills any spores, bacteria, and microorganisms, leaving a sterile and clean product.

Rice hulls are also often used as part of a mix of growing media such as 30%-40% rice hulls and pine bark mix. The overall pH of parboiled and composted rice hulls range from 5.7 to 6.5, which is right in the pH range for most hydroponically grown plants. Fresh and/or composted rice hulls tend to have a high Manganese (Mn) content. But problems with Manganese toxicity can be avoided as long as the pH is above 5. Which is below normal range for hydroponics anyway.

References

- Hydroponics: explainthatstuff.com, Retrieved 28 April, 2020
- Hydroponic-farming, irrigation-irrigation-systems: novagric.com, Retrieved 18 June, 2020
- Hydroponic-systems: urbanagriculturereview.wordpress.com, Retrieved 14 March, 2020
- Organic-hydroponics-3211: maximumyield.com, Retrieved 26 July, 2020
- Hydroponic-production-systems-impact-on-nutritional-status-and-bioactive-compounds-of-fresh-vegetabl, vegetables-importance-of-quality-vegetables-to-human-health: intechopen.com, Retrieved 19 January, 2020
- Automation-and-robotics-used-in-hydroponic-system: intechopen.com, Retrieved 17 July, 2020
- Mediums-page, mediums: homehydrosystems.com, Retrieved 13 June, 2020

2

Methods and Techniques used in Hydroponics

There are various methods and techniques that are used in hydroponics such as aeroponics, fogponics, anthroponics, aquaponics, rotary hydroponics, kratky method, vertical farming, etc. This chapter discusses the different methods and techniques used in hydroponics for a thorough understanding of the subject.

Static Solution Culture

Static solution culture is a method by which plants are full-grown in reservoirs filled with an artificially made nutrient solution. Containers can be plastic buckets, mason jars, tanks, tubs and so on. Containers are generally made out of plastic; however other matcrials can also be used. The solution used should be aerated but can also be left unaerated. If the solution is kept aerated, then it is kept at a low level so that the roots of the plant can get sufficient oxygen to breathe. An opening is cut into the brim of the container for every plant to come out. A single container can accommodate numerous plants, provided the container is large enough according to the growth size of the plant.

If this procedure is being carried on at home then the method can be carried on in plastic food buckets or glass cans with aeration supplied by a normal aquarium airline tube, an aquarium pump or an aquarium valve. Containers should be covered with butcher paper, thick black plastic, other aluminum foils or any other material that can keep away light which can prevent the growth of algae.

This nutrient liquid is changed according to a schedule or once a week when its concentration level jumps down beyond a definite level which can be identified with electrical conductivity indicator. Each time the concentration falls down, either fresh nutrient liquid or water is again added. A float valve or a Mariotte's bottle can be used to repeatedly uphold the solution concentration level.

Raft solution system necessitates the plants to be placed on a sheet of floating plastic which is then floated on the surface of this nutrient liquid. This way the solution point does not drop beyond the root level.

Hydroponics has always been considered as one of the most effective alternative to traditional mode of cultivation, and the rapid increase of this method has brought in high yields from countries around the world. This method allows the common crowd to adapt this technology in their homes or private properties. A method like this always brings in higher advantages like the elimination of soil and the cost of water and nutrients. A control method like this will always bring in more nutrition in plants and cause less health hazards caused by pesticides and other chemicals.

Static hydroponic systems can then be divided into two large categories, open and closed systems. In open systems, the nutrient solution that is given to the plants is never recovered and is "wasted" while in closed systems the solution is used for as long as it has the right chemical properties to feed the plants. For example, a lettuce raft system with no water recirculation is considered a static closed system (since the solution stays in contact with the plant until it is not right anymore) while a drip irrigation system with no pumps is considered an open system because new solution is given to the plants continuously.

Open Static Hydroponic System

The largest advantage of open static hydroponic systems is that the cost and infrastructure needed to pump solution back into the nutrient reservoir is neglected, this cost can be significant if solution volumes are small. Another important advantage is that the concentration of nutrients and pH of the nutrient solution does not need to be checked because fresh solution is continuously provided.

The main drawbacks of these systems have to do with the cost of the nutrient solution spent and the dumping of hydroponic nutrient solution in either soil, rivers, or the sewage system. Hydroponic nutrient solutions are very contaminating because of the easiness with which they cause algae blooms. The solution needs to be correctly processed in order to guarantee no contamination occurs, doing this will most often require water pumps, something which makes the first advantages of these systems void. The cost of the nutrient solution also becomes prohibitive. A regular plant consumes about half a gallon per hour in a drip irrigation system, wasting this amount of solution every hour in a commercial facility is unacceptable (reason why no commercial growers use open static hydroponic systems).

Closed Static Hydroponic System

These systems have the advantage of not being contaminating and using nutrient solutions as effectively as possible. They are the most economical systems that can be built and they provide ideal growing conditions for most plants with short life cycles. Larger plants like tomatoes do not benefit from having their roots wet all the time and require some sort of special design within the system.

The main disadvantages of these systems have to do with the necessity for water oxygenation (using air pumps) and the need for continuous monitoring of the nutrient solution in order to guarantee that ideal conditions are always met. This of course, is not very complicated and can be done with no problems. Most commercial growers will couple the ideal of a static system with a dynamic one in order to further improve its effectiveness. For example, commercial lettuce raft growers often add water pumps in order to circulate, sterilize and oxygenate the water instead of using air pumps to do this.

Most static hydroponic systems are pretty primitive and are often confined to small home growers whose concern with efficiency and optimal conditions is not as strict as those of commercial growers. Almost all commercial growers use dynamic systems of some sort and that should also be the aim of home growers as these systems have shown to be more efficient and ecological albeit more expensive than static cultivation methods.

Continuous-flow Solution Culture

In hydroponics, a continous-flow solution culture have a solution that is constantly moving past the roots. Thus, they roots are not submerged in standing water. For this type of set-up, it is common to use big storage tanks rather than individual jars or buckets for the plants.

With this method the nutrient solution perpetually flows past the roots. The continuous flow solution culture is easier to automate than the static solution culture, because sampling and adjustments to the temperature and nutrient concentration can be made in large supply tanks that are designed for thousands of plants. A very popular variation is the so called "nutrient film technique", a technique where a very shallow stream of water containing all the dissolved nutrients requiered for plant growth is re-circulated past the bare roots of the plants in a watertight channel.

In an ideal system, the depth of the re-circulating stream should be very shallow so that the top of the thick root mat, which develops in the bottom of the channel, although moist, is in the air. As a result, an abundant supply of oxygen is provided to the roots of the plants. NFT, because of its design, provides a system wherein all three requirements for healthy plant growth can be met at the same time - provided that the simple concept of NFT is always remembered and practiced. The result of these advantages is that higher yields of high-quality produce are obtained over an extended period of cropping. A downside of NFT is that it has very little buffering against interruptions in the flow, e.g., power outages, but, overall, it is one of the more productive techniques. The same design characteristics apply to all conventional NFT systems.While slopes along channels of 1:100 have been recommended, in practice it is difficult to build a base for channels that is sufficiently true to enable nutrient films to flow without ponding in locally depressed areas. As a consequence, it is recommended that slopes of 1:30 to 1:40 be used. This allows for minor irregularities in the surface, but, even with these slopes, ponding and water logging may occur.

The slope may be provided by the floor, or benches or racks may hold the channels and provide the required slope. Both methods are used and depend on local requirements, often determined by the site and crop requirements. As a general guide, flow rates for each gully should be 1 litre per minute. At planting, rates may be half this and the upper limit of 2L/min appear about the maximum. Flow rates beyond these extremes are often associated with nutritional problems. Depressed growth rates of many crops have been observed when channels exceed 12 metres in length.On rapidly growing crops, tests have indicated that, while oxygen levels remain adequate, nitrogen may be depleted over the length of the gully. As a consequence, channel length should not exceed 10 15 metres. In situations where this is not possible, the reductions in growth can be eliminated by placing another nutrient feed halfway along the gully and reducing flow rates to 1L/min through each outlet.

Deep Water Culture

Deep water culture (DWC) is a hydroponic method of plant production by means of suspending the plant roots in a solution of nutrient-rich, oxygenated water. Also known as raft/pond or float systems, this method uses floating rafts to suspend plant roots into a pond of water often 8-12 inches deep. Compared to other hydroponic techniques, such as NFT, deep water culture is relatively inexpensive to set up and can be easily reproduced by a home grower. Since there is a relatively large reservoir of nutrient rich water for each plant, there is buffering for pH, EC, and temperature, meaning those elements of the systems won't fluctuate as fast as they might in an NFT system.

Hobby Methods

Most hobby growers use Deep Water Culture (DWC). Net pots, plastic pots with netting to allow roots to grow through their surface, are filled with a hydroponic medium such as Hydroton or Rockwool to hold the base of the plant. In some cases net pots are not needed. For Oxygenation of the hydroponic solution, an airstone is added. This air stone is then connected to an airline that runs to an air pump.

As the plant grows, the root mass stretches through the rockwool or hydroton into the water below. Under ideal growing conditions, plants are able to grow a root mass that comprises the entire bin in a loosely packed mass. As the plant grows and consumes nutrients the pH and EC of the water fluctuate. For this reason, frequent monitoring must be kept of the nutrient solution to ensure that it remains in the uptake range of the crop. A pH that is too high or too low will make certain nutrients unavailable for uptake by plants. Generally, the best pH for hydroponic crops is around 5.5-6.0. In terms of EC, too low means that there is a low salt content, usually meaning a lack of fertilizer, and an EC that it too high indicates a salt content that could damage the roots of crops. Desired EC depends on the crop that is growing. A common EC for leafy greens is somewhere between 1.5-2.2.

Recirculation Deep Water Culture

Traditional methods using unconnected buckets require each bucket to be tested for pH and conductivity factor (CF) individually. This has led to the creation of Recirculation Deep Water Culture (RDWC) systems. Rather than having individual buckets, RDWC bins are linked together most commonly using a PVC pipe. A pump is also added at the front of the system that pulls water through a line from rear of the system into a control bucket. This return line generally has a spin filter on it that cleans particulate from the water before it reaches the pump. The individual bins, including the control are aerated. The primary disadvantage of rDWC is that disease can spread quickly in these systems which can facilitate the transfer of pathogens from one reservoir to another.

Commercial Deep Water Culture

In a commercial system, there is usually a large pond where crops float on a raft. Seedlings are germinated in cubes (such as rockwool, oasis, or other media) and then transplanted into the rafts. Plants may be re-spaced during the growth period (higher density at first, and lower density later). The nutrient solution is oxygenated through air pumps or recirculation, and water is chilled to a temperature between 18-24C in order to maintain proper dissolved oxygen concentration, which is crucial to plant growth. Chilling the water also helps to prevent pathogens such as pythium, and delay bolting. PH (optimum 5.5-6.0) and EC (dependent on crop) are controlled with acid or base injectors and fertilizer injectors, respectively. Supplemental lighting can be added to make sure that the plants receive the correct amount of light. Lighting depends on the crop, and the stage of growth for the crop. For example, lettuce grows best with 15-17 mol·m-2·d-1 of light.

Typically only short-statured crops such as leafy greens and herbs are grown commercially in deep water culture as rafts move through the pond and taller plants (such as tomatoes or cucumbers) would necessitate being trellised. The most common commercial deep water culture crop is lettuce. Lettuce does best in a pH of 5.6-6.0, EC of 1.1-1.2 (of fertilizer), 17 $mol{\cdot}m^{-2}{\cdot}d_{-1}$ daily light integral which may consist of a combination of natural and supplemental lighting, air temperature of 24 °C day/19 °C night, water temperature of 25 °C, and dissolved oxygen of 7 $mg{\cdot}L^{-3}$.

The rafts are cleaned after each harvest by scrubbing to remove organic matter and applying bleach or other sanitizing agents. Aeroponics.

Lettuce and wheat grown in an aeroponic apparatus.

Aeroponics is the process of growing plants in an air or mist environment without the use of soil or an aggregate medium. Aeroponic culture differs from both conventional hydroponics, aquaponics, and in-vitro (plant tissue culture) growing. Unlike hydroponics, which uses a liquid nutrient solution as a growing medium and essential minerals to sustain plant growth; or aquaponics which uses water and fish waste, aeroponics is conducted without a growing medium. It is sometimes considered a type of hydroponics, since water is used in aeroponics to transmit nutrients.

In 1911, V.M.Artsikhovski published in the journal "Experienced Agronomy" an article "On Air Plant Cultures", which talks about his method of physiological studies of root systems by spraying various substances in the surrounding air - the aeroponics method. He designed the first aeroponics and in practice showed their suitability for plant cultivation.

It was W. Carter in 1942 who first researched air culture growing and described a method of growing plants in water vapor to facilitate examination of roots. As of 2006, aeroponics is used in agriculture around the globe.

In 1944, L.J. Klotz was the first to discover vapor misted citrus plants in a facilitated research of his studies of diseases of citrus and avocado roots. In 1952, G.F. Trowel grew apple trees in a spray culture.

It was F. W. Went in 1957 who first coined the air-growing process as "aeroponics", growing coffee plants and tomatoes with air-suspended roots and applying a nutrient mist to the root section.

GTi's Genesis Rooting System.

The first commercially available aeroponic apparatus was manufactured and marketed by GTi in 1983. It was known then as the Genesis Machine - taken from the movie Star Trek II: The Wrath of Khan. The Genesis Machine was marketed as the "Genesis Rooting System".

GTi's device incorporated an open-loop water driven apparatus, controlled by a microchip, and delivered a high pressure, hydro-atomized nutrient spray inside an aeroponic chamber. The Genesis Machine connected to a water faucet and an electrical outlet.

Aeroponic Propagation

GTi's apparatus cut-away of vegetative cutting propagated aeroponically.

Aeroponic culturing revolutionized cloning (vegetative propagation) from cuttings of plants. Numerous plants which were previously considered difficult, or impossible, became easier to propagate via stem cuttings in aeroponics, such as delicate hardwoods or cacti which were sensitive to bacterial infection in cuttings. The overall success of propagation with the use of aeroponics is that the system creates a highly aerated environment around the root, which causes good root hair development. There is also more root and growth development due to the nutrients supplied to the plants through the aeroponics system. Since the roots are not grown in any rooting media, it minimizes the risk of the plants getting infected by root disease.

The use of aeroponic is important for helping to propagate plants with low rate of success in vegetative propagation, plants that have important medicinal uses, plants that are in high demand, and to create new cultivars of certain plant species. Leptadenia reticulata is an important plant used in medicines that also has a low reproduction rate through both seed and cuttings. Aeroponics has made it easier to propagate some these important medicinal plants. Ulmus Americana, which was almost completely wiped out by Dutch elm disease, along with other cultivars of the species have also shown some success through propagation with aeroponics allowing for elm trees to be more available on the market.

Aeroponics is a more viable alternative to the traditionally used process of overhead misters. There is a higher success rate with the use of aeroponics compared to overhead misters, and with the overhead misters there are drawbacks such as needing to apply large volumes of water, having potentially unsanitary conditions, having irregular misting coverage, and potential leaching of foliar nutrients. In short, cloning became easier because the aeroponic apparatus initiated faster and cleaner root development through a sterile, nutrient rich, highly oxygenated, and moist environment.

Air-rooted Transplants

Cloned aeroponics transplanted directly into soil.

Aeroponics significantly advanced tissue culture technology. It cloned plants in less time and reduced numerous labor steps associated with tissue culture techniques. Aeroponics could eliminate stage I and stage II plantings into soil (the bane of all tissue culture growers). Tissue culture plants must be planted in a sterile media (stage-I) and expanded out for eventual transfer into sterile soil (stage-II). After they are strong enough they are transplanted directly to field soil. Besides being labor-intensive, the entire process of tissue culture is prone to disease, infection, and failure.

With the use of aeroponics, growers cloned and transplanted air-rooted plants directly into field soil. Aeroponic roots were not susceptible to wilting and leaf loss, or loss due to transplant shock (something hydroponics can never overcome). Because of their healthiness, air-rooted plants were less likely to be infected with pathogens. (If the RH of the root chamber gets above 70 degrees F, fungus gnats, algae, anaerobic bacteria are likely to develop).

The efforts by GTi ushered in a new era of artificial life support for plants capable of growing naturally without the use of soil or hydroponics. GTi received a patent for an all-plastic aeroponic method and apparatus, controlled by a microprocessor in 1985.

Aeroponics became known as a time and cost saver. The economic factors of aeroponic's contributions to agriculture were taking shape.

Genesis Growing System

GTi's Aeroponic Growing System greenhouse facility.

By 1985, GTi introduced second generation aeroponics hardware, known as the "Genesis Growing System". This second generation aeroponic apparatus was a closed-loop system. It utilized recycled effluent precisely controlled by a microprocessor. Aeroponics graduated to the capability of supporting seed germination, thus making GTi's the world's first plant and harvest aeroponic system.

Many of these open-loop unit and closed-loop aeroponic systems are still in operation today.

Commercialization

Aeroponics eventually left the laboratories and entered into the commercial cultivation arena. In 1966, commercial aeroponic pioneer B. Briggs succeeded in inducing roots on hardwood cuttings by air-rooting. Briggs discovered that air-rooted cuttings were tougher and more hardened than those formed in soil and concluded that the basic principle of air-rooting is sound. He discovered air-rooted trees could be transplanted to soil without suffering from transplant shock or setback to normal growth. Transplant shock is normally observed in hydroponic transplants.

In Israel in 1982, L. Nir developed a patent for an aeroponic apparatus using compressed low-pressure air to deliver a nutrient solution to suspended plants, held by styrofoam, inside large metal containers.

In summer 1976, British researcher John Prewer carried out a series of aeroponic experiments near Newport, Isle of Wight, U.K., in which lettuces (variety Tom Thumb) were grown from seed to maturity in 22 days in polyethylene film tubes made rigid by pressurized air supplied by ventilating fans. The equipment used to convert the water-nutrient into fog droplets was supplied by Mee Industries of California. "In 1984 in association with John Prewer, a commercial grower on the Isle of Wight - Kings Nurseries - used a different design of aeroponics system to grow strawberry plants. The plants flourished and produced a heavy crop of strawberries which were picked by the nursery's customers. The system proved particularly popular with elderly customers who appreciated the cleanliness, quality and flavor of the strawberries, and the fact they did not have to stoop when picking the fruit".

In 1983, R. Stoner filed a patent for the first microprocessor interface to deliver tap water and nutrients into an enclosed aeroponic chamber made of plastic. Stoner has gone on to develop numerous companies researching and advancing aeroponic hardware, interfaces, biocontrols and components for commercial aeroponic crop production.

The first commercial aeroponic greenhouse for aeroponic food production.

In 1985, Stoner's company, GTi, was the first company to manufacture, market and apply large-scale closed-loop aeroponic systems into greenhouses for commercial crop production.

In the 1990s, GHE or General Hydroponics [Europe] thought to try to introduce aeroponics to the hobby hydroponics market and finally came to the Aerogarden system. However, this could not be classed as 'true' aeroponics because the Aerogarden produced tiny droplets of solution rather than a fine mist of solution; the fine mist was meant to reproduce true Amazon rain. In any case, a product was introduced to the market and the grower could broadly claim to be growing their hydroponic produce aeroponically. A demand for aeroponic growing in the hobby market had been established and moreover it was thought of as the ultimate hydroponic growing technique. The difference between true aeroponic mist growing and aeroponic droplet growing had become very blurred in the eyes of many people. At the end of the nineties, a UK firm, Nutriculture, was encouraged enough by industry talk to trial true aeroponic growing; although these trials showed positive results compared with more traditional growing techniques such as NFT and Ebb & Flood there were drawbacks, namely cost and maintenance. To accomplish true mist aeroponics a special pump had to be used which also presented scalability problems. Droplet-aeroponics was easier to manufacture, and as it produced comparable results to mist-aeroponics, Nutriculture began development of a scalable, easy to use droplet-aeroponic system. Through trials they found that aeroponics was ideal for plant propagation; plants could be propagated without medium and could even be grown-on. In the end, Nutriculture acknowledged that better results could be achieved if the plant was propagated in their branded X-stream aeroponic propagator and moved on to a specially designed droplet-aeroponic growing system - the Amazon.

Aeroponically Grown Food

In 1986, Stoner became the first person to market fresh aeroponically grown food to a national grocery chain. He was interviewed on NPR and discussed the importance of the water conservation features of aeroponics for both modern agriculture and space.

Contemporary Aeroponics

Contemporary aeroponic techniques have been researched at NASA's research and commercialization center BioServe Space Technologies located on the campus of the University of Colorado in Boulder, Colorado. Other research includes enclosed loop system research at Ames Research Center, where scientists were studying methods of growing food crops in low gravity situations for future space colonization.

In 2000, Stoner was granted a patent for an organic disease control biocontrol technology that allows for pesticide-free natural growing in an aeroponic systems.

In 2004, Ed Harwood, founder of AeroFarms, invented an aeroponic system that grows lettuces on micro fleece cloth. AeroFarms, utilizing Harwood's patented aeroponic technology, is now operating the largest indoor vertical farm in the world based on annual growing capacity in Newark, New Jersey. By using aeroponic technology the farm is able to produce and sell up to two million pounds of pesticide-free leafy greens per year.

Aeroponic Bio-pharming

Aeroponically grown biopharma corn.

Aeroponic bio-pharming is used to grow pharmaceutical medicine inside of plants. The technology allows for completed containment of allow effluents and by-products of biopharma crops to remain inside a closed-loop facility. As recently as 2005, GMO research at South Dakota State University by Dr. Neil Reese applied aeroponics to grow genetically modified corn.

According to Reese it is a historical feat to grow corn in an aeroponic apparatus for bio-massing.

The university's past attempts to grow all types of corn using hydroponics ended in failure.

Using advanced aeroponics techniques to grow genetically modified corn Reese harvested full ears of corn, while containing the corn pollen and spent effluent water and preventing them from entering the environment. Containment of these by-products ensures the environment remains safe from GMO contamination.

Reese says, aeroponics offers the ability to make bio-pharming economically practical.

Large Scale Integration of Aeroponics

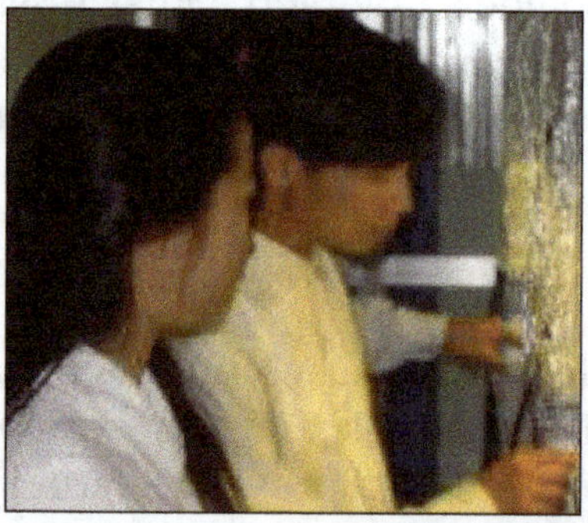

Aeroponic Graduate Program: Hanoi Agricultural University.

In 2006, the Institute of Biotechnology at Vietnam National University of Agriculture, in joint efforts with Stoner, established a postgraduate doctoral program in aeroponics. The university's Agrobiotech Research Center, under the direction of Professor Nguyen Quang Thach, is using aeroponic laboratories to advance Vietnam's minituber potato production for certified seed potato production.

Aeroponic potato explants on day 3 after insertion in the aeroponic system.

The historical significance for aeroponics is that it is the first time a nation has specifically called out for aeroponics to further an agricultural sector, stimulate farm economic goals, meet increased demands, improve food quality and increase production.

"We have shown that aeroponics, more than any other form of agricultural technology, will significantly improve Vietnam's potato production. We have very little tillable land, aeroponics makes complete economic sense to us", attested Thach.

Aeroponic greenhouse for potato minituber product.

Vietnam joined the World Trade Organization (WTO) in January 2007. The impact of aeroponics in Vietnam will be felt at the farm level.

Aeroponic integration in Vietnamese agriculture will begin by producing a low cost certified disease-free organic minitubers, which in turn will be supplied to local farmers for their field plantings of seed potatoes and commercial potatoes. Potato farmers will benefit from aeroponics because their seed potatoes will be disease-free and grown without pesticides. Most importantly for the Vietnamese farmer, it will lower their cost of operation and increase their yields, says Thach.

Methods

The basic principle of aeroponic growing is to grow plants suspended in a closed or semi-closed environment by spraying the plant's dangling roots and lower stem with an atomized or sprayed, nutrient-rich water solution. The leaves and crown, often called the canopy, extend above. The roots of the plant are separated by the plant support structure. Often, closed-cell foam is compressed around the lower stem and inserted into an opening in the aeroponic chamber, which decreases labor and expense; for larger plants, trellising is used to suspend the weight of vegetation and fruit.

Ideally, the environment is kept free from pests and disease so that the plants may grow healthier and more quickly than plants grown in a medium. However, since most aeroponic environments are not perfectly closed off to the outside, pests and disease may still cause a threat. Controlled environments advance plant development, health, growth, flowering and fruiting for any given plant species and cultivars.

Due to the sensitivity of root systems, aeroponics is often combined with conventional hydroponics, which is used as an emergency "crop saver" backup nutrition and water supply if the aeroponic apparatus fails.

High-pressure aeroponics is defined as delivering nutrients to the roots via 20–50 micrometre mist heads using a high-pressure (80 pounds per square inch (550 kPa)) diaphragm pump.

Benefits and Drawbacks

Increased Air Exposure

Many types of plants can be grown aeroponically.

Air cultures optimize access to air for successful plant growth. Materials and devices which hold and support the aeroponic grown plants must be devoid of disease or pathogens. A distinction of a true aeroponic culture and apparatus is that it provides plant support features that are minimal. Minimal contact between a plant and support structure allows for 100% of the plant to be entirely in air. Long-term aeroponic cultivation requires the root systems to be free of constraints surrounding the stem and root systems. Physical contact is minimized so that it does not hinder natural growth and root expansion or access to pure water, air exchange and disease-free conditions.

Close-up of the first patented aeroponic plant support structure. Its unrestricted support of the plant allows for normal growth in the air/moisture environment, and is still in use today.

Benefits of Oxygen in the Root Zone

Oxygen (O_2) in the rhizosphere (root zone) is necessary for healthy plant growth. As aeroponics is conducted in air combined with micro-droplets of water, almost any plant can grow to maturity in air with a plentiful supply of oxygen, water and nutrients.

Some growers favor aeroponic systems over other methods of hydroponics because the increased aeration of nutrient solution delivers more oxygen to plant roots, stimulating growth and helping to prevent pathogen formation.

Clean air supplies oxygen which is an excellent purifier for plants and the aeroponic environment. For natural growth to occur, the plant must have unrestricted access to air. Plants must be allowed to grow in a natural manner for successful physiological development. The more confining the plant support becomes, the greater incidence of increasing disease pressure of the plant and the aeroponic system.

Some researchers have used aeroponics to study the effects of root zone gas composition on plant performance. Soffer and Burger studied the effects of dissolved oxygen concentrations on the formation of adventitious roots in what they termed "aero-hydroponics". They utilized a 3-tier hydro and aero system, in which three separate zones were formed within the root area. The ends of the roots were submerged in the nutrient reservoir, while the middle of the root section received nutrient mist and the upper portion was above the mist. Their results showed that dissolved O_2 is essential to root formation, but went on to show that for the three O_2 concentrations tested, the number of roots and root length were always greater in the central misted section than either the submersed section or the un-misted section. Even at the lowest concentration, the misted section rooted successfully.

Other Benefits of Air (CO_2)

Plants in a true aeroponic apparatus have 100% access to the CO_2 concentrations ranging from 450 ppm to 780 ppm for photosynthesis. At one mile (1.6 km) above sea level, the CO_2 concentration in the air is 450 ppm during daylight. At night, the CO_2 level will rise to 780 ppm. Lower elevations will have higher levels. In any case, the air culture apparatus offers the ability for plants to have full access to all of the available CO_2 in the air for photosynthesis. Growing under lights during the evening allows aeroponics to benefit from the natural occurrence.

Disease-free Cultivation

Aeroponics can limit disease transmission since plant-to-plant contact is reduced and each spray pulse can be sterile. In the case of soil, aggregate, or other media, disease can spread throughout the growth media, infecting many plants. In most greenhouses, these solid media require sterilization after each crop and, in many cases, they are simply discarded and replaced with fresh, sterile media.

A distinct advantage of aeroponic technology is that if a particular plant does become diseased, it can be quickly removed from the plant support structure without disrupting or infecting the other plants.

Due to the disease-free environment that is unique to aeroponics, many plants can grow at higher density (plants per square meter) when compared to more traditional forms of cultivation (hydroponics, soil and Nutrient Film Technique [NFT]). Commercial aeroponic systems incorporate hardware features that accommodate the crop's expanding root systems.

Basil grown from seed in an aeroponic system located inside a modern greenhouse.

Researchers have described aeroponics as a "valuable, simple, and rapid method for preliminary screening of genotypes for resistance to specific seedling blight or root rot".

The isolating nature of the aeroponic system allowed them to avoid the complications encountered when studying these infections in soil culture.

Water and Nutrient Hydro-atomization

Aeroponic equipment involves the use of sprayers, misters, foggers, or other devices to create a fine mist of solution to deliver nutrients to plant roots. Aeroponic systems are normally closed-looped systems providing macro and micro-environments suitable to sustain a reliable, constant air culture. Numerous inventions have been developed to facilitate aeroponic spraying and misting. The key to root development in an aeroponic environment is the size of the water droplet. In commercial applications, a hydro-atomizing spray at 360° is employed to cover large areas of roots utilizing air pressure misting.

A variation of the mist technique employs the use of ultrasonic foggers to mist nutrient solutions in low-pressure aeroponic devices.

Water droplet size is crucial for sustaining aeroponic growth. Too large a water droplet means less oxygen is available to the root system. Too fine a water droplet, such as those generated by the ultrasonic mister, produce excessive root hair without developing a lateral root system for sustained growth in an aeroponic system.

Mineralization of the ultrasonic transducers requires maintenance and potential for component failure. This is also a shortcoming of metal spray jets and misters. Restricted access to the water causes the plant to lose turgidity and wilt.

Advanced Materials

NASA has funded research and development of new advanced materials to improve aeroponic reliability and maintenance reduction. It also has determined that high pressure hydro-atomized mist of 5-50 micrometres micro-droplets is necessary for long-term aeroponic growing.

For long-term growing, the mist system must have significant pressure to force the mist into the dense root systems. Repeatability is the key to aeroponics and includes the hydro-atomized droplet size. Degradation of the spray due to mineralization of mist heads inhibits the delivery of the water nutrient solution, leading to an environmental imbalance in the air culture environment.

Special low-mass polymer materials were developed and are used to eliminate mineralization in next generation hydro-atomizing misting and spray jets.

Nutrient Uptake

Close-up of roots grown from wheat seed using aeroponics.

The discrete nature of interval and duration aeroponics allows the measurement of nutrient uptake over time under varying conditions. Barak et al. used an aeroponic system for non-destructive measurement of water and ion uptake rates for cranberries.

In their study, these researchers found that by measuring the concentrations and volumes of input and efflux solutions, they could accurately calculate the nutrient uptake rate (which was verified by comparing the results with N-isotope measurements). After verification of their analytical method, Barak et al. went on to generate additional data specific to the cranberry, such as diurnal variation in nutrient uptake, correlation between ammonium uptake and proton efflux, and the relationship between ion concentration and uptake. Work such as this not only shows the promise of aeroponics as a research tool for nutrient uptake, but also opens up possibilities for the monitoring of plant health and optimization of crops grown in closed environments.

Atomization (>65 pounds per square inch (450 kPa)), increases bioavailability of nutrients, consequently, nutrient strength must be significantly reduced or leaf and root burn will develop. Note the large water droplets in the photo to the right. This is caused by the feed cycle being too long or the pause cycle too short; either discourages both lateral root growth and root hair development. Plant growth and fruiting times are significantly shortened when feed cycles are as short as possible. Ideally, roots should never be more than slightly damp nor overly dry. A typical feed/pause cycle is < 2 seconds on, followed by ~1.5-2 minute pause- 24/7, however, when an accumulator system is incorporated, cycle times can be further reduced to < ~1 second on, ~1 minute pause.

As a Research Tool

Soon after its development, aeroponics took hold as a valuable research tool. Aeroponics offered researchers a noninvasive way to examine roots under development. This new technology also allowed researchers a larger number and a wider range of experimental parameters to use in their work.

The ability to precisely control the root zone moisture levels and the amount of water delivered makes aeroponics ideally suited for the study of water stress. K. Hubick evaluated aeroponics as a means to produce consistent, minimally water-stressed plants for use in drought or flood physiology experiments.

Aeroponics is the ideal tool for the study of root morphology. The absence of aggregates offers researchers easy access to the entire, intact root structure without the damage that can be caused by removal of roots from soils or aggregates. It's been noted that aeroponics produces more normal root systems than hydroponics.

Types of Aeroponics

Low-pressure Units

In most low-pressure aeroponic gardens, the plant roots are suspended above a reservoir of nutrient solution or inside a channel connected to a reservoir. A low-pressure pump delivers nutrient solution via jets or by ultrasonic transducers, which then drips or drains back into the reservoir. As plants grow to maturity in these units they tend to suffer from dry sections of the root systems, which prevent adequate nutrient uptake. These units, because of cost, lack features to purify the nutrient solution, and adequately remove incontinuities, debris, and unwanted pathogens. Such units are usually suitable for bench top growing and demonstrating the principles of aeroponics.

High-pressure Devices

High-pressure aeroponic techniques, where the mist is generated by high-pressure pumps, are typically used in the cultivation of high value crops and plant specimens that can offset the high setup costs associated with this method of horticulture.

High-pressure aeroponics systems include technologies for air and water purification, nutrient sterilization, low-mass polymers and pressurized nutrient delivery systems.

Commercial Systems

Commercial aeroponic systems comprise high-pressure device hardware and biological systems. The biological systems matrix includes enhancements for extended plant life and crop maturation.

Biological subsystems and hardware components include effluent controls systems, disease prevention, pathogen resistance features, precision timing and nutrient solution pressurization, heating and cooling sensors, thermal control of solutions, efficient photon-flux light arrays, spectrum filtration spanning, fail-safe sensors and protection, reduced maintenance & labor saving features, and ergonomics and long-term reliability features.

Commercial aeroponic systems, like the high-pressure devices, are used for the cultivation of high value crops where multiple crop rotations are achieved on an ongoing commercial basis.

Advanced commercial systems include data gathering, monitoring, analytical feedback and internet connections to various subsystems.

Aeroponics in Space

Space Plants

NASA life support GAP technology with untreated beans (left tube) and biocontrol treated beans (right tube) returned from the Mir space station aboard the space shuttle.

Plants were first taken into Earth's orbit in 1960 on two separate missions, Sputnik 4 and Discoverer 17. On the former mission, wheat, pea, maize, spring onion, and Nigella damascena seeds were carried into space, and on the latter mission Chlorella pyrenoidosa cells were brought into orbit.

Plant experiments were later performed on a variety of Bangladesh, China, and joint Soviet-American missions, including Biosatellite II (Biosatellite program), Skylab 3 and 4, Apollo-Soyuz, Sputnik, Vostok, and Zond. Some of the earliest research results showed the effect of low gravity on the orientation of roots and shoots.

Subsequent research went on to investigate the effect of low gravity on plants at the organismic, cellular, and subcellular levels. At the organismic level, for example, a variety of species, including pine, oat, mung bean, lettuce, cress, and Arabidopsis thaliana, showed decreased seedling, root, and shoot growth in low gravity, whereas lettuce grown on Cosmos showed the opposite effect of growth in space (Halstead and Scott 1990). Mineral uptake seems also to be affected in plants grown in space. For example, peas grown in space exhibited increased levels of phosphorus and potassium and decreased levels of the divalent cations calcium, magnesium, manganese, zinc, and iron.

Biocontrols in Space

In 1996, NASA sponsored Stoner's research for a natural liquid biocontrol, known then as ODC (organic disease control), that activates plants to grow without the need for pesticides as a means to control pathogens in a closed-loop culture system. ODC is derived from natural aquatic materials.

By 1997, Stoner's biocontrol experiments were conducted by NASA. BioServe Space Technologies's GAP technology (miniature growth chambers) delivered the ODC solution unto bean seeds. Triplicate ODC experiments were conducted in GAP's flown to the MIR by the space shuttle; at the Kennedy Space Center; and at Colorado State University (J. Linden). All GAPS were housed in total darkness to eliminate light as an experiment variable. The NASA experiment was to study only the benefits of the biocontrol.

NASA's experiments aboard the MIR space station and shuttle confirmed that ODC elicited increased germination rate, better sprouting, increased growth and natural plant disease mechanisms when applied to beans in an enclosed environment. ODC is now a standard for pesticide-free aeroponic growing and organic farming. Soil and hydroponics growers can benefit by incorporating ODC into their planting techniques. ODC meets USDA NOP standards for organic farms.

Aeroponics for Space and Earth

In 1998, Stoner received NASA funding to develop a high performance aeroponic system for earth and space. Stoner demonstrated that a dry bio-mass of lettuce can be significantly increased with aeroponics. NASA utilized numerous aeroponic advancements developed by Stoner. Due to this advancement we can use as a reference to space aeroponics.

Abstract: The purpose of the research conducted was to identify and demonstrate technologies for high-performance plant growth in a variety of gravitational environments. A low-gravity environment, for example, poses the problems of effectively bringing water and other nutrients to the plants and effecting recovery of effluents. Food production in the low-gravity environment of space provides further challenges, such as minimization of water use, water handling, and system weight. Food production on planetary bodies such as the Moon or Mars also requires dealing with a hypogravity environment. Because of the impacts to fluid dynamics in these various gravity environments, the nutrient delivery system has been a major focus in plant growth system optimization.

NASA aeroponic lettuce seed germination. Day 30.

There are a number of methods currently utilized (both in low gravity and on Earth) to deliver nutrients to plants. Substrate dependent methods include traditional soil cultivation, zeoponics, agar, and nutrient-loaded ion exchange resins. In addition to substrate dependent cultivation, many methods using no soil have been developed such as nutrient film technique, ebb and flow, aeroponics, and many other variants. Many hydroponic systems can provide high plant performance but nutrient solution throughput is high, necessitating large water volumes and substantial recycling of solutions, and the control of the solution in hypogravity conditions is difficult at best.

Aeroponics, with its use of a hydro-atomized spray to deliver nutrients, minimizes water use, increases oxygenation of roots, and offers excellent plant growth, while at the same time approaching or bettering the low nutrient solution throughput of other systems developed to operate in low gravity. Aeroponics' elimination of substrates and the need for large nutrient stockpiles reduces the amount of waste materials to be processed by other life support systems. Furthermore, the absence of substrates simplifies planting and harvesting (providing opportunities for automation), decreases the volume and weight of expendable materials, and eliminates a pathway for pathogen transmission. These many advantages combined with the results of this research that prove the viability of aeroponics in microgravity makes aeroponics a logical choice for efficient food production in space.

NASA Inflatable Aeroponics

In 1999, Stoner, funded by NASA, developed an inflatable low-mass aeroponic system (AIS) for space and earth for high performance food production.This advancements are very useful in space aeroponics.

Abstract: Aeroponics International's (AI's) innovation is a self-contained, self-supporting, inflatable aeroponic crop production unit with integral environmental systems for the control and delivery of a nutrient/mist to the roots. This inflatable aeroponic system addresses the needs of subtopic 08.03 Spacecraft Life Support Infrastructure and, in particular, water and nutrient delivery systems technologies for food production. The inflatable nature of our innovation makes it lightweight, allowing it to be deflated so it takes up less volume during transportation and storage. It improves on AI's current aeroponic system design that uses rigid structures, which use more expensive materials, manufacture processes, and transportation. As a stationary aeroponic system, these existing high-mass units perform very well, but transporting and storing them can be problematic.

On Earth, these problems may hinder the economic feasibility of aeroponics for commercial growers. However, such problems become insurmountable obstacles for using these systems on long-duration space missions because of the high cost of payload volume and mass during launch and transit.

The NASA efforts lead to developments of numerous advanced materials for aeroponics for earth and space.

Benefits of Aeroponics for Earth and Space

NASA aeroponic lettuce seed germination- Day 3.

Aeroponics possesses many characteristics that make it an effective and efficient means of growing plants.

Less Nutrient Solution Throughout

Plants grown using aeroponics spend 99.98% of their time in air and 0.02% in direct contact with hydro-atomized nutrient solution. The time spent without water allows the roots to capture oxygen more efficiently. Furthermore, the hydro-atomized mist also significantly contributes to the effective oxygenation of the roots. For example, NFT has a nutrient throughput of 1 liter per minute compared to aeroponics' throughput of 1.5 milliliters per minute.

NASA aeroponic lettuce seed germination- Day 12.

The reduced volume of nutrient throughput results in reduced amounts of nutrients required for plant development.

Another benefit of the reduced throughput, of major significance for space-based use, is the reduction in water volume used. This reduction in water volume throughput corresponds with a reduced buffer volume, both of which significantly lighten the weight needed to maintain plant growth. In addition, the volume of effluent from the plants is also reduced with aeroponics, reducing the amount of water that needs to be treated before reuse.

The relatively low solution volumes used in aeroponics, coupled with the minimal amount of time that the roots are exposed to the hydro-atomized mist, minimizes root-to-root contact and spread of pathogens between plants.

Greater Control of Plant Environment

NASA aeroponic lettuce seed germination (close-up of root zone environment)- Day 19.

Aeroponics allows more control of the environment around the root zone, as, unlike other plant growth systems, the plant roots are not constantly surrounded by some medium (as, for example, with hydroponics, where the roots are constantly immersed in water).

Improved Nutrient Feeding

A variety of different nutrient solutions can be administered to the root zone using aeroponics without needing to flush out any solution or matrix in which the roots had previously been immersed. This elevated level of control would be useful when researching the effect of a varied regimen of nutrient application to the roots of a plant species of interest. In a similar manner, aeroponics allows a greater range of growth conditions than other nutrient delivery systems. The interval and duration of the nutrient spray, for example, can be very finely attuned to the needs of a specific plant species. The aerial tissue can be subjected to a completely different environment from that of the roots.

More User-friendly

The design of an aeroponic system allows ease of working with the plants. This results from the separation of the plants from each other, and the fact that the plants are suspended in air and the roots are not entrapped in any kind of matrix. Consequently, the harvesting of individual plants is quite simple and straightforward. Likewise, removal of any plant that may be infected with some type of pathogen is easily accomplished without risk of uprooting or contaminating nearby plants.

More Cost Effective

Close-up of aeroponically grown corn and roots inside an aeroponic (air-culture) apparatus.

Aeroponic systems are more cost effective than other systems. Because of the reduced volume of solution throughput (discussed above), less water and fewer nutrients are needed in the system at any given time compared to other nutrient delivery systems. The need for substrates is also eliminated, as is the need for many moving parts.

Use of Seed Stocks

With aeroponics, the deleterious effects of seed stocks that are infected with pathogens can be minimized. This is due to the separation of the plants and the lack of shared growth matrix. In addition, due to the enclosed, controlled environment, aeroponics can be an ideal growth system in which to grow seed stocks that are pathogen-free. The enclosing of the growth chamber, in addition to the isolation of the plants from each other helps to both prevent initial contamination from pathogens introduced from the external environment and minimize the spread from one plant to others of any pathogens that may exist.

Fogponics

Fogponics is an offshoot of aeroponics. In this indoor gardening practice, plants are suspended in enclosed systems with no soil or growing medium. Instead, a fog or mist of water and nutrients is pumped into the closed system to give the roots constant exposure to the nutrients the plant needs to grow.

The theory behind fogponics is that plants are best able to absorb particulate nutrients in the 1 to 25 micrometer (μm) range. Therefore, plants in a fog or mist mixture can better absorb nutrients for more efficient gardening and less waste.

Fogponics is an advancement on aeroponic gardening techniques. With fogponics, though, an opaque crate, bucket, or tray is used to seal in the plants' roots. Holes are cut in the top to accommodate the net cups in which the plants will grow, and one more hole is cut in the side or bottom for an electric fogger.

The fogger introduces a mist of nutrient-water mixture into the closed system, and the plants' roots are constantly exposed to nutrients at particulate sizes that they easily absorb and take up.

Because there is no need for soil, fogponics offers a sustainable and efficient means to grow vegetables and other plants indoors. Fogponics is also more efficient than hydroponics in that all the water vapor and nutrients for the system are trapped within the system and nothing is lost to evaporation.

Fogponics does require regular use of electric power to introduce a mist or fog into the system. When setting up a fogponics system, indoor horticulturalists must consider a backup power supply during a power outage, as it could be devastating for their plants. Likewise, the system must be regularly maintained to effectively deliver nutrients.

Fogponics, however, allows growers to create very specific nutrient-water mixtures to create more suitable growing conditions for certain plants. Some vegetables, such as tomatoes, respond very well to fogponics. Furthermore, with fogponics it is easier to evenly distribute and disperse nutrients and water to all roots equally.

Fogponics is a form of aeroponics, sometimes called the next phase in aeroponic technology. Think of it as a refinement, but a significant one. It uses the same basic concept as aeroponics the practice of suspending plant roots in the air of a grow box and supplying water/nutrient droplets on a timed schedule. In fogponics, though, the droplets are so tiny, they qualify as vapor. And in this case, smaller is better.

Plants absorb elements most efficiently in the range of about 1-25 micrometers (μm), and though foggers work somewhat differently depending on the manufacturer and model, a fogging system will supply vapor at between 5 and 30 μm.

While aeroponic systems supply nutrients to plant roots, foggers deliver vapor to stems and leaf nodes as well as roots, resulting in more efficient nutrient uptake and healthier plants overall.

Plants absorb droplets well, but because they absorb vapor better and faster, this can be good news for growers.

There's less waste, good space conservation and less opportunity for error, especially during root feeding.

How can you make a mistake spritzing plant roots to feed them? Aeroponic systems rely on precisely spaced nozzles to spray the interior of a grow box evenly.

Sometimes there are voids, or plant roots can become so thick they create obstructions.

Clogged nozzles can also fail, killing plants that are locked out of water and nutrients. These types of problems may not be detected until it's too late.

However, with fogging systems, wider and more comprehensive dispersion is much easier to accomplish, and growers are more likely to get to all the roots evenly and consistently.

This type of delivery system may also change the nature of the roots on some plants by encouraging smaller, hair-like growths that are even better at taking advantage of vapor-based feeding.

In a basic hydroponic set-up, media is a substitute for soil. There are lots of different types of hydroponic media, but they all tend to be chemically inert, heavy enough to anchor plant roots and porous enough to allow for good oxygen access and water/nutrient distribution.

There are some drawbacks to using media, though. Media can clog feed lines, entangle roots, harbor bacteria, algae or fungi, or lose porosity over time.

In some systems, switching from one type of media to another can be a problem, too.

Immersion Systems

In an immersion system, delicate roots remain in a liquid environment most if not all of the time, and definitely throughout a plant's early development.

The anchoring media is replaced by a basket or other simple support structure that makes roots easy to see and easier to remove when or if plants are eventually transferred to soil or another growing environment.

When roots are suspended in water more than intermittently, they need special treatment in the form of supplemental oxygen.

The gaps and pores in conventional growing media make oxygen readily available to roots, and conventional hydroponic systems also include processes that remove and replace water and nutrients regularly, giving plants increased exposure to the air.

In submersion systems, oxygen is added through the use of an air stone, the same way oxygenated water is supplied to fish in an aquarium.

Because plants are in the water/nutrient mixture all of the time, pH fluctuations and temperature instabilities can happen suddenly and be devastating.

Going Airborne

Aeroponics approaches things differently by bringing water/nutrients to plant roots suspended in air instead of providing oxygen to plant roots suspended in water. The basic idea is pretty simple. Plants are installed on a platform using a support collar.

The roots dangle below into a box structure containing a network of spray nozzles that deliver nutrients to them on a timed schedule. Excess nutrients are recycled and reused.

During the early development of aeroponics, some interesting things came to light. One of the most significant is that the increased access to oxygen in an aeroponic system enhances plant growth a lot.

Another is that the continuous delivery of nutrients and oxygen in lavish amounts is an efficient way to farm.

Plants don't waste energy searching for food, and since there is no real competition for nutrients, and no incidental reduction in oxygen supply, plants can be grown in close proximity to one another, saving space, time and money.

The fogponic system uses electric foggers to pump and vibrate under pressure to transform nutrient and water mix into humidity, like the humid atmosphere found in the rainforest. This creates a constant humid and nutrient-rich fog for plant roots.

With fogponics, the root system can have a full coverage as the tiny droplet size of the gravity-defying fog can travel and penetrate all space around it.

How does it Work?

Like any other active hydroponics system, plants are given with what they need, moisture, nutrients, oxygen. They do not have to actively seek for these like in the soil counterparts. All of their efforts are spent on growing, flowering, or sprouting (for seedlings), and root development (for cloning).

But unlike traditional Hydroponics, where plants get the oxygen from the water oxygenated by an air pump, Fogponics/Aeroponics works by hanging the plants in the air. It is held by a tray/foam. The plant roots are suspended, so they have plenty of oxygen access. Water is provided by plants via the foggers. The action is usually automated by a timer.

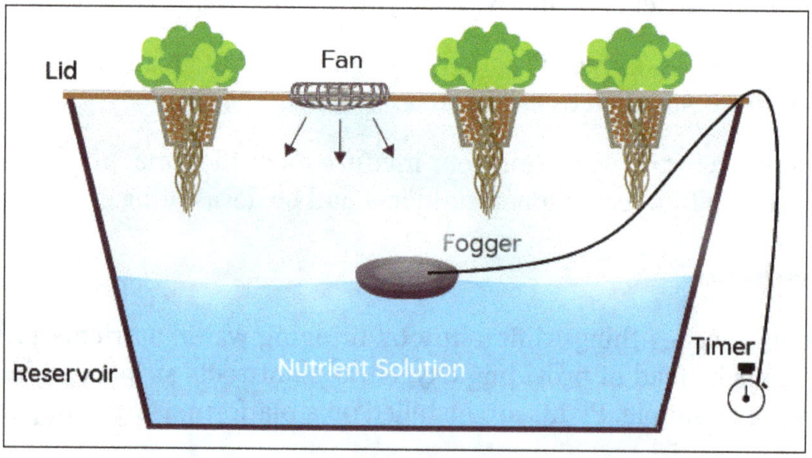

Fogponics operates just like aeroponics. But instead using the water misting, the foggers are used.They produce and atomize much smaller droplets than in aeroponics, usually less than 10 microns in size.

How to Set up a Fogponic System?

You'll need these tools to set up the system.

A container, a reservoir, a bucket or anything large enough that can hold water and at least 1 plant.

A lid. This will cover your reservoir and keep the fog from escaping. You'll also need to drill some holes to place the net pot with the plants in it.

A net pot. Get the size you find fit with your system. Fill it with the growing media (hydroton, perlite, vermiculite). Place the net pot on the reservoir. It must be held by the lid or styrofoam so that when plants are put into the net pot, their roots are hung above the water of the reservoir.

Fogponic Hydroponic nutrient. Your plants do need nutrients, right? The liquid form is recommended as it works best in fogponics systems. You can also use the powder form, just make sure it is totally dissolved. Stay away from any organic nutrient form because it can clog your fogponic system. After choosing the nutrients, Fill them in the reservoir.

The ultrasonic fogger (This is the most important apparatus as it decides whether you call it aeroponics or fogponics). Put it in the reservoir. Turn it on, and let the fog come out.

A micro fan (optional). You can place it above or below the root. Aim the fan at the direction that helps regulate the fog around the roots.

pH measurement tools. Get a pH meter to first check the pH of the system, and occasionally recheck it later.

A pH control kit. We prefer pH up and down. If the pH of our system is too low or too high, just add the appropriate pH control kit into it.

What Plants Work Best with the Fogponics

In theory, you can grow all plants with the Fogponics, yet it's advised that you choose appropriate plants for the best results.

Seedling and Clonings

There is an excellent reason why fogponics works absolutely well with seedlings and clonings. Young cloners from cuttings do not have a developed root systems yet, so it's really difficult to moisture and provide nutrients to them. The amounts of water must be kept at an appropriate level. Because too much that can suffocate the cloning. Too little, or the roots may dry out, and sprouts will never spring up.

With the fogs emitted from the fogponics systems, growers provide a constant mix of small weak moisture for cloners at controlled amounts.

Green Vegs

The fogponics also works best with green vegs, including lettuces, spinaches, kales, cucumbers, beans and so on.

Herbs

Undoubtedly, most herbs like basil, mint, chives, etc thrive in the fogponic system as it is lightweight, and have a short growing life span.

Pros of a Fogponic System

Tiny Sized Droplets and Great Coverage

According to the NASA research - Aeroponics for Spaceflight Plant Growth by J.M. Clawson, A. Hoehn, L.S. Stodieck and P. Todd BioServe Space Technologies R.J. Stoner 2000, small droplets tend not to impinge on roots. That answers why aeroponics/fogponics is so effective for clonings because the cuttings, and seedlings are still weak, and these low-pressure fog will not hurt the cuttings and roots.

Also, the hanging roots will get more air than they could in the soil, and in traditional Hydroponic systems like Deep Water Culture, Nutrient Film Technique.

What's more, Fogponics does not deliver much nutrient as the small-sized nature of the fog, which is very important for Seedlings as they do not need many nutrients.

High Nutrient Concentration

Because of this, the nutrient concentration is not lost when traveling within the reservoir. To see how it is a benefit, let's compare with a Hydroponic system, say the Nutrient Film Technique. In the NFT system, when the nutrient cycle comes and runs across the plant roots, those at the beginning of the tray will get the most nutrient, and it becomes less concentrated at the end of the water cycle.

Easy to Clean

The fogponics only has 1 unit to clean rather than lots of heads like in the Aeroponics.

Cons of a Fogponic System

Heats from the Atomizer

When running continuously for a long time, the atomizer will warm the reservoir. The heat can dry and evaporate the fogs. And your roots can dry out subsequently. If that happens, you need to the cool the temperature down. You can do this by getting a timer and set on and off interval for the foggers. Another way is using some chilling methods like adding some ices into it. If you have more money to spend, you can get a water chiller.

Built up Salt

Like the aeroponics system, over time you'll see the salt accumulates in the system. This can clog the foggers. That means it'll need regularly cleaning by using a toothbrush or soak them in vinegar to ensure the fogponics will run effectively.

Susceptible to Power Outage

As the advanced form of hydroponics, the fogponics relies on the electricity to run the whole system. Also, the plant's roots are hung freely without submerging into the water like the traditional Hydroponic systems. In this sense, in case the electricity is off, the fog will stop. The plants will not able to get the moisture and nutrients. They will dry out fast and the death can be expected.

High Initial Cost

This advanced form of Hydroponics will cost you some start-up cost in the beginning.

Growing with Fogponic Tip

Because of the heats and the risk of evaporation of the fog, you can use a small fan to regulate the fogs.

Occasionally check the nutrients to ensure the salt is not built up as it will hinder your plants' development.

Most of the time, 1 single-node fogger is enough for starters. But when the sizes of your gardens extend, you'll need to get more.

When growing with seedling/cloners. As plants do not need not a large amount of nutrients at this phase, don't add many nutrients into the reservoir.

Wrapping Up

Even though there are some hurdles like some startup cost, regular maintenance, risks from the system interruptions, etc, Fogponics is still a very effective Hydroponic system.

It is fairly interesting to try for Hydroponic starters and hobbyists.

In commercial greenhouses, growers have used aerponics/fogponics for growing plants for years but they will use more expensive misting and foggers for better fog production instead of the common cheap ultrasonic foggers.

Anthroponics

Anthroponics can be defined as a recirculating soilless agriculture system that uses natural bacterial cycles to convert human biowaste, such as urine, into plant fertilizer. Anthroponic systems will typically combine aspects of aquaponic farming, organic hydroponics, and wastewater treatment.

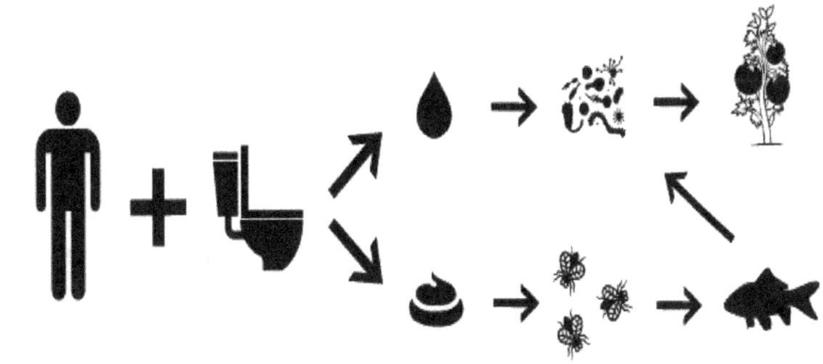

Anthroponics: Waste to Food Cycle

As a soiless system, anthroponics reaps the benets of hydroponic farming by using less water and recirculating it continuously. When compared to traditional agriculture, hydroponic agriculture uses ten times less water. At the same time, it is a more sustainable form of hydroponics as the source of the nutrients is a waste product (of humans). Most hydroponic systems use nutrients sourced from mines, requiring fossil fuel in its extraction, preparation and transportation. Human urine can instead be locally sourced and prepared for anthroponic use without the energy requirements of more common hydroponic mineral nutrient solutions.

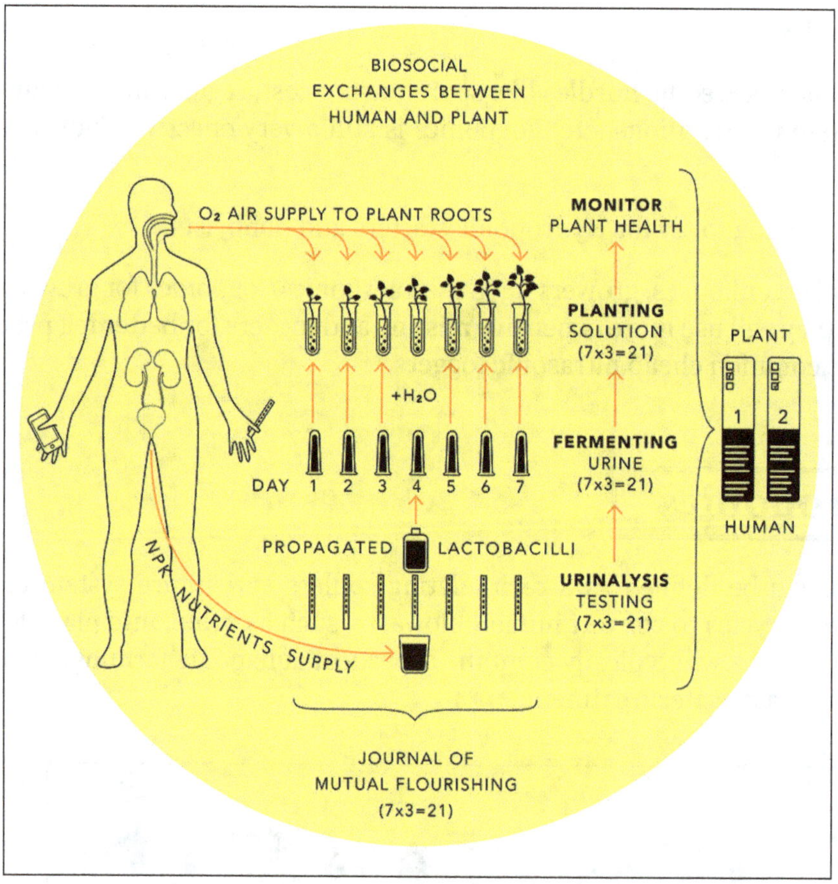

While there is some debate whether urine is sterile or not, it is understood that letting the urine sit in a sealed container will let the urea convert to carbamic acid and then un-ionized ammonia, which will raise the pH of the solution to above 9, eectively sterilizing it through the reduction of the bacterial population is process is known as ageing, and can be accelerated by adding urease enzyme to the solution.

An anthroponics system should therefore take into account both the urine collection, the ageing process, the nitrication, and the hydroponic component for easy incorporation in existing infrastructure and scalability. One example of an anthroponics process can be seen.

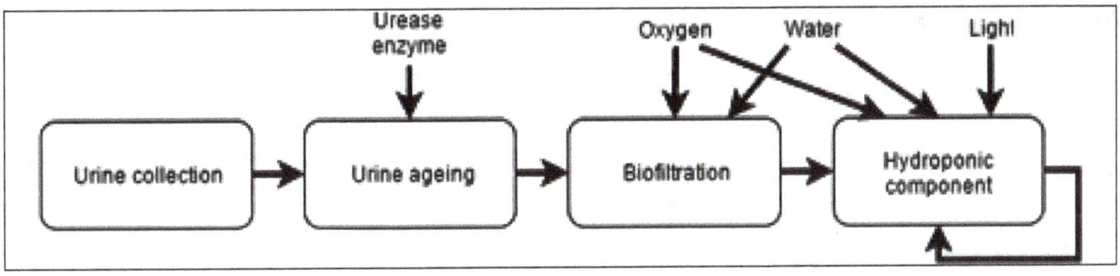

In order to move to the implementation of anthroponics, several smallscale experiments were performed to test proof-of-concepts, dimensioning, ratios, and optimization for more nutrient demanding crops.

Aquaponics

Aquaponics is a combination of aquaculture, which is growing fish and other aquatic animals, and hydroponics which is growing plants without soil. Aquaponics uses these two in a symbiotic combination in which plants are fed the aquatic animals' discharge or waste. In return, the vegetables clean the water that goes back to the fish. Along with the fish and their waste, microbes play an important role to the nutrition of the plants. These beneficial bacteria gather in the spaces between the roots of the plant and converts the fish waste and the solids into substances the plants can use to grow. The result is a perfect collaboration between aquaculture and gardening.

Aquaponics is a big hope for sustainable organic crop production, aquaculture and water consumption. The fish waste is recycled and used for plant growth instead of throwing it in the ocean. The water is recirculated in a closed system lowering the consumption of this resource.

Types of Systems

Since aquaponics uses basically the same systems as hydroponics, there aren't many differences in how the system works, except for the added fish in the water tanks. Drip irrigation, flood and drain, deep culture or water submerged roots, and nutrient film technique are highly compatible and customizable to merge with growing fish.

Importance of pH Control in Aquaponics

pH is an important part of aquaculture. Setting it to a perfect level can be a bit confusing since there are three living organisms to care for: your plants, your fish, and the bacteria inside the water and each of them has a different pH need. A neutral pH from 6.8 to 7.2 is good for the aquaponic garden. Because of the fish waste, the pH will become acidic and you will need to use aquaponic compatible pH adjusters. If the pH level is not beneficial for the system that is too low or too high, the plants will not be able to absorb nutrients optimally and your fish will die eventually. It's very important to monitor the pH level each day and to keep it within the neutral range.

A too alkaline or too acidic pH is one of the main reasons fish or plants die, leading to gardening failure. The pH adjusters need to be specially designed for this type of growing system, otherwise, they could harm the fish. You can find these adjusters in a local aquaponic gardening supplier. Another thing to keep in mind is the water hardness because it affects how pH will behave when trying to adjust it. Sometimes it would be necessary to also take care of the water hardness when working the pH. Fish don't like sudden changes in pH, so when adjusting it try to lower or increase it slowly.

Fish and other Aquatic Animals you can Grow in Aquaponics

Fish are the ones feeding your plants. The fish used in this type of aquaculture are freshwater fish, most popular being tilapia and barramundi because they tolerate better diverse water conditions and they grow fast. Trout can also be used especially for lower water temperatures. Other aquatic animals you can grow are snails and shrimps.

You can feed the fish special food you can purchase in an animal store or other foods like water lettuce and duckweed.

Vegetables that Grow in Aquaponics

In an a small aquaponic based garden you can grow vegetables that don't need heavy nutrient input. Lettuce, kale, watercress, arugula, decorative flowers, mint, herbs, okras, spring onions and leek, radishes, spinach and other small vegetables. Cabbage, tomatoes, cucumbers, beans, broccoli and cauliflower can require more nutrition and a well-stocked or more advanced aquaponic system. Avoid growing plants that need acidic or alkaline water, because those levels of pH can definitely harm the fish.

Benefits of aquaponics:

- Aquaponics is a way to grow your own fish and vegetables at the same time. You feed the fish and the fish will feed your plants through their waste output.
- There is no need to use fertilizers because the fish provide rich nutrients for the plants.
- In aquaponics, less water is used for the crops. Research has shown that aquaponic gardens use 1/10th of the water you would use for soil garden.
- Regular gardening pesticides or other chemicals can't be used because they would harm the fish.
- This results in healthier and organic vegetables.
- You won't experience any soil borne diseases in aquaponics because there is no soil.
- You can grow plants in very small space, and have a great harvest.
- Plants grow fast because they get very nutritious substances from the fish waste.
- Plants and fish production can be done in a controlled temperature environment.
- Water is used in a closed system and circulated effectively, reducing the consumption and the water bills.

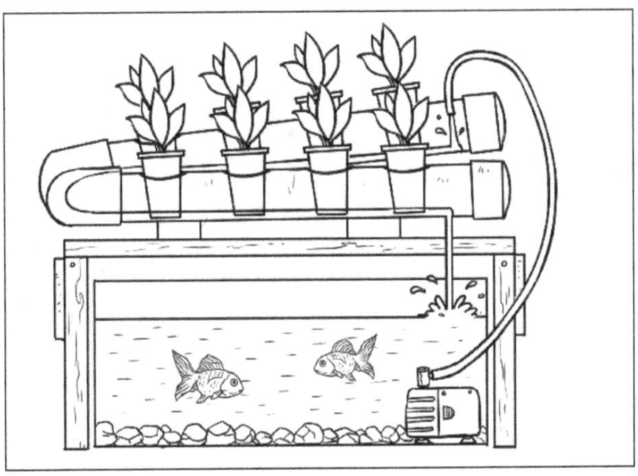

Setting up an Aquaponic Garden at Home

- You can make your own aquaponics system and here is a simple and complete guide you can use to make one.

- Have set a different power source as a backup. It's vital to keep the water flowing and the oxygen pumps on.
- Make sure you feed the fish enough and let them thrive. Depletion of fish stock makes this type of cultivation impossible.
- Keep food input constant for the fish and that will result in regular fish waste you can use to feed your plants.
- Ensure your plants and fish with good aeration. Not only the plants need their roots to be oxygenated, but also, the fish and the bacteria need the water to be oxygenated. As the fish grow bigger, their oxygen needs increases and you might need to adjust accordingly.
- When you decide what plants you want to grow, pick the ones that have similar water condition needs as the fish, and you will have greater success.
- Remove some excess fish waste when necessary. Too much can harm the health of the fish.
- Keep an eye on the level of pH because, it is crucial for the garden.
- Fish tanks should be made of glass or food grade plastic.
- Avoid using any pesticide other than organic, or any other substances that can and will harm the fish or the good bacteria (vinegar, citric and/or hydrochloric acid).

Passive Hydroponics

Cattleya transferred to passive hydroponics culture 5 weeks earlier. Rich development of surface roots.

Passive hydroponics, semi-hydroponics or passive subirrigation is a method of growing plants without soil, peat moss, or bark. Instead an inert porous medium transports water and fertilizer to the roots by capillary action. Water and fertilizer are held in a reservoir and conducted to the roots as necessary, reducing labor and providing a constant supply of water to the roots. In the simplest method, the pot sits in a shallow solution of fertilizer and water or on a capillary mat saturated with nutrient solution. Since routine maintenance is much simplified, passive hydroponics can reduce the labor required to maintain a large collection of plants. The Kratky Method is a specialized passive hydroponic technique which involves plants suspended in net pots above a non-circulating water and nutrient reservoir.

The various hydroponic media available contain more air space than more traditional potting mixes, delivering increased oxygen to the roots. Allowing air to reach the roots is particularly important in preventing root rot in epiphytic plants such as orchids and bromeliads, whose roots are exposed to the air in nature. Passive hydroponics also may add additional ambient humidity through evaporation. It is important in passive subirrigation to wash out the system from time to time to remove salt accumulation. Plants that require drying between waterings or a dry dormant period may fail to thrive under the constant moisture of passive hydroponics.

Media

The medium is a key component. It must must provide capillary action that will keep the plant uniformly moist. It has to be something that won't decompose and provides adequate air space. Many media are available for passive hydroponics, but the most common are expanded clay pellets, coconut husk chips, perlite, vermiculite, diatomite, and rock wool. These are frequently used in combination.

Disadvantages of Passive Sub-irrigation

Plants growing in hydroponic systems either grow with their roots directly in water or, as in the case of passive sub-irrigation, in moisture-retaining materials such as fiberglass, clay pebbles, coconut husk or perlite. In passive sub-irrigation, the inert medium acts as a wick to carry water from a reservoir below the planting up to the plant's roots. Like conventional hydroponic growing, the water contains all the nutrients the plants need.

Greenhouse Growing

In passive sub-irrigation, plants are grown in porous media that transport water and nutrients to roots. Water is run past the media through a system of pipes, and sits in the bottom of plant-containing trays. Sub-irrigation systems usually exist in controlled environments, because the rate at which nutrients are mixed into the solution is very important. Therefore using sub-irrigation systems outdoors is often difficult, so if you do not have a good indoor space for setting up a sub-irrigation system, it may not be for you.

Expensive Equipment

Another drawback to passive sub-irrigation is that it requires special equipment, the cost of which can be prohibitive when you are first starting up. Although costs vary considerably depending on the type of equipment you buy and how large a setup you desire, it is certainly more expensive than simply planting in the ground. If you already have an indoor hydroponics system, then retrofitting it to become a passive sub-irrigation system may prove less expensive. This means modifying your system so that, instead of suspending plants in water, they are only periodically exposed to it.

Fertilizer Frequency

A disadvantage of all hydroponics systems, including passive sub-irrigation, compared to traditional growing techniques, is the frequency with which you must supply the plants with nutrients. Because they do not have access to soil, which contains many of the minerals and nutrients lacking in water-based growing systems, you must supply it to them in the water. Buying these nutrients is an extra expense over the intermittent fertilization you might give garden plants, though as passive sub-irrigation does recirculate its water, unnecessary loss of nutrients is prevented.

Disease Development

Because passive sub-irrigation waters plants from below, relying on the roots' capillary action to draw liquid up into the plant, most plants grown in this way are significantly less prone to foliage diseases. However, when insect or disease problems do take root, it is often more difficult to manage them because you cannot simply apply drenches the way you would in the garden. Therefore keeping an eye out for disease is very important, and using integrated pest management strategies becomes even more crucial.

Rotary Hydroponics

Rotary hydroponics is a style of farming that was initially developed by NASA for use on the International Space Station. Now, a brilliant new company called Bace, founded in 2018, has created a rotary hydroponics system for people's homes called Rotofarm.

It is similar to all hydroponic systems in the sense that it works without soil and uses mess-free nutrient and water reservoirs to feed its plants. However, everything else about it is new and futuristic, like an evolved version of the traditional hydroponic system.

It's made from stainless steel, silicone and strengthened, smoked glass. Yet, as beautiful and mesmerizing as it is, "It was paramount that Rotofarm not be treated as decor," said its creator. "It's designed to be used. We went to great lengths to ensure it fit seamlessly into any kitchen". This wonderful hydroponic system puts a whole garden on your countertop but it's not just a patch of plants, it is a work of art.

The founder of Bace, Australian Toby Farmer, refined the principles of rotary hydroponics, resulting in the manifestation of Rotofarm, a beautiful device that he calls the "easiest-to-use and most automated planter in the world". Farmer believes "rotary hydroponics is the answer to growing vegetables faster, easier and more efficiently".

What makes rotary hydroponics so much better? Its rotating design turns the plants in a full circle every hour meaning that half the plants spend their time in negative gravity. According to Bace, this accelerates their growth as much as twice over. Furthermore, since the plants can grow all along the surface of the circle, people can enjoy nearly 5 feet (1.5 meters) of an indoor vegetable patch while only using up only 11 inches (30 centimeters) of counter space.

Special features of the design include an automated LED light tube at the center of the farm-bed that shines out onto the tops of all the plants and can be adjusted differently in each quadrant. What that means is that the light tube is divided into four zones that can be dimmed and timed differently to provide the optimal level of light for different plants.

The color of the light is similar to sunlight, but with a hint more red to increase the rate of photosynthesis, stimulating even faster plant growth. To eliminate the issue of glare in the house from the bright light emitted by the rotary garden, there is an optional magnetic cover that reduces the brightness while still allowing the plants inside to be visible.

While testing the prototype, Farmer's approach to eating and cooking had changed for the better. It's also a much more sustainable option than store-bought vegetables. He said.

By growing produce at home without preservatives or pesticides, the produce doesn't expire until its cut, you're not spending as much money on groceries and a lot of space is saved in your refrigerator. Perhaps most importantly, we help to reduce carbon emissions produced by mass farming, transport and plastic packaging into supermarkets.

Rotofarm is a "truly set-and-forget" machine as it can even notify the user when produce has overgrown and requires harvesting.

Kratky Method

The Kratky method is a passive hydroponic technique for growing plants suspended above a reservoir of nutrient-rich water. Because it is a non-circulating technique, no additional inputs of water or nutrients are needed after the original application, and no electricity, pumps, or water and oxygen circulation systems are required. The Kratky method has applications both for commercial food production and as a small-scale and low-maintenance technique for home growers.

A 2" net pot, which can be used as part of a Kratky system.

Plants are placed in net cups filled with an inert growth medium such as rock wool, hydroton, or coconut coir. The net cups are suspended above a reservoir of water containing essential nutrients in solution. Only the root tips are allowed to touch the surface of the reservoir. As the plant grows and depletes the water level, a gap of moist air will form and expand between the water surface and the base of the plant. The roots in this gap become laterally branching "oxygen roots," and absorb oxygen from the air inside the container. By the time the water level is fully depleted, the plant should be ready to harvest. Thus, in one growing cycle, no additional replenishment of water or nutrients is needed beyond the initial application.

Three lettuce plants grown using the Kratky method in mason jars.

The method is named after Bernard Kratky, a researcher at the University of Hawaii who first proposed the method in the journal Actae Horticulturae in 2009. Kratky discusses using floating pallets for commercial food production. Multiple plants are placed in net pots on boards that float atop a shared reservoir, with support beams slightly underneath the water level. As water is depleted, the boards eventually come to rest on the support beams, providing an air gap to allow oxygen intake. The Kratky method has also found popularity among home growers, and is often carried out on smaller scales in containers such as buckets and mason jars. The technique is designed for leafy vegetables that do not consume large amounts of water, so the water supply does not need to be continually replenished.

Vertical Farming

The practice of growing crops in vertically stacked layers.

Vertical farming is the practice of growing crops in vertically stacked layers. It often incorporates controlled-environment agriculture, which aims to optimize plant growth, and soilless farming techniques such as hydroponics, aquaponics, and aeroponics. Some common choices of structures to house vertical farming systems include buildings, shipping containers, tunnels, and abandoned mine shafts.

Lettuce grown in indoor vertical farming system.

The modern concept of vertical farming was proposed in 1999 by Dickson Despommier, professor of Public and Environmental Health at Columbia University. Despommier and his students came up with a design of a skyscraper farm that could feed 50,000 people. Although the design has not yet been built, it successfully popularized the idea of vertical farming. Current applications of vertical farmings coupled with other state-of-the-art technologies, such as specialized LED lights, have resulted in over 10 times the crop yield than would receive through traditional farming methods. There have been several different means of implementing vertical farming systems into communities such as: Paignton, Israel, Singapore, Chicago, Munich, London, Japan, and Lincolnshire.

The main advantage of utilizing vertical farming technologies is the increased crop yield that comes with a smaller unit area of land requirement. The increased ability to cultivate a larger variety of crops at once because crops do not share the same plots of land while growing is another sought-after advantage. Additionally, crops are resistant to weather disruptions because of their placement indoors, meaning less crops lost to extreme or unexpected weather occurrences. Lastly, because of its limited land usage, vertical farming is less disruptive to the native plants and animals, leading to further conservation of the local flora and fauna.

Vertical farming technologies face economic challenges with large start-up costs compared to traditional farms. In Victoria, Australia, a "hypothetical 10 level vertical farm" would cost over 850 times more per cubic meter of arable land than a traditional farm in rural Victoria. Vertical farms also face large energy demands due to the use of supplementary light like LEDs. Moreover, if non-renewable energy is used to meet these energy demands, vertical farms could produce more pollution than traditional farms or greenhouses.

Techniques of Vertical Farming

Hydroponics

Hydroponics refers to the technique of growing plants without soil. In hydroponic systems, the roots of plants are submerged in liquid solutions containing macronutrients, such as nitrogen, phosphorus, sulphur, potassium, calcium, and magnesium, as well as trace elements, including iron, chlorine, manganese, boron, zinc, copper, and molybdenum. Additionally, inert (chemically inactive) mediums such as gravel, sand, and sawdust are used as soil substitutes to provide support for the roots.

Indoor Hydroponics of Morus.

The advantages of hydroponics include the ability to increase yield per area and reduce water usage. A study has shown that, compared to conventional farming, hydroponic farming could increase the yield per area of lettuce by around 11 times while requiring 13 times less water. Due to these advantages, hydroponics is the predominant growing system used in vertical farming.

Aquaponics with catfish.

Aquaponics

The term aquaponics is coined by combining two words: aquaculture, which refers to fish farming, and hydroponics the technique of growing plants without soil. Aquaponics takes hydroponics one step further by integrating the production of terrestrial plants with the production of aquatic organisms in a closed-loop system that mimics nature itself. Nutrient-rich wastewater from the fish tanks is filtered by a solid removal unit and then led to a bio-filter, where toxic ammonia is converted to nutritious nitrate. While absorbing nutrients, the plants then purify the wastewater, which is recycled back to the fish tanks. Moreover, the plants consume carbon dioxide produced by the fish, and water in the fish tanks obtains heat and helps the greenhouse maintain temperature at night to save energy. As most commercial vertical farming systems focus on producing a few fast-growing vegetable crops, aquaponics, which also includes an aquacultural component, is currently not as widely used as conventional hydroponics.

Aeroponics

Aeroponically-grown chives.

The invention of aeroponics was motivated by the initiative of NASA (the National Aeronautical and Space Administration) to find an efficient way to grow plants in space in the 1990s. Unlike conventional hydroponics and aquaponics, aeroponics does not require any liquid or solid medium to grow plants in. Instead, a liquid solution with nutrients is misted in air chambers where the plants are suspended. By far, aeroponics is the most sustainable soil-less growing techniques, as it uses up to 90% less water than the most efficient conventional hydroponic systems and requires no replacement of growing medium. Moreover, the absence of growing medium allows aeroponic systems to adopt a vertical design, which further saves energy as gravity automatically drains away excess liquid, whereas conventional horizontal hydroponic systems often require water pumps for controlling excess solution. Currently, aeroponic systems have not been widely applied to vertical farming, but are starting to attract significant attention.

Controlled-environment Agriculture

Controlled-environment agriculture (CEA) is the modification of the natural environment to increase crop yield or extend the growing season. CEA systems are typically hosted in enclosed structures such as greenhouses or buildings, where control can be imposed on environmental factors including air, temperature, light, water, humidity, carbon dioxide, and plant nutrition. In vertical farming systems, CEA is often used in conjunction with soilless farming techniques such as hydroponics, aquaponics, and aeroponics.

Types of Vertical Farming

Building-based Vertical Farms

Abandoned buildings are often reused for vertical farming, such as a farm at Chicago called "The Plant," which was transformed from an old meatpacking plant. However, new builds are sometimes also constructed to house vertical farming systems. For example, a company named "Vertical Harvest" built a three-story hydroponic greenhouse next to a parking lot in Jackson, Wyoming, and aims to grow 100,000 lbs of produce annually.

Shipping-container Vertical Farms

Recycled shipping containers are an increasingly popular option for housing vertical farming systems. The shipping containers serve as standardized, modular chambers for growing a variety of plants, and are often equipped with LED lighting, vertically stacked hydroponics, smart climate controls, and monitoring systems. Moreover, by stacking the shipping containers, farms can save space even further and achieve higher yield per square foot. Currently, there are many commercial shipping-container vertical-farming units on the market, such as the "Greenery" from Freight Farms and the "TerraFarm" from Local Roots.

Deep Farms

A "deep farm" is a vertical farm built from refurbished underground tunnels or abandoned mine shafts. As temperature and humidity underground are generally temperate and constant, deep farms require less energy for heating. Deep farms can also use nearby groundwater to reduce the cost of water supply. Despite low costs, a deep farm can produce 7 to 9 times more food than a conventional farm above ground on the same area of land, according to Saffa Riffat, chair in Sustainable Energy at the University of Nottingham. Coupled with automated harvesting systems, these underground farms can be fully self-sufficient.

A company named "Growing Underground" claims to have built the world's first underground farm, and is growing greens in a refurbished World War II bomb shelter 33 meters under Clapham, London. Their products are available in local supermarkets such as Whole Foods, Planet Organic, and M&S.

Proposal of Vertical Farming

Dickson Despommier, professor of Public and Environmental Health at Columbia University, founded the root of the concept of vertical farming. In 1999, he challenged his class of graduate students to calculate how much food they could grow on the rooftops of New York. The student concluded that they could only feed about 1000 people. Unsatisfied with the results, Despommier suggested growing plants indoors instead, on multiple layers vertically. Despommier and his students then proposed a design of a 30-story vertical farm equipped with artificial lighting, advanced hydroponics, and aeroponics that could produce enough food for 50,000 people. They further outlined that approximately 100 kinds of fruits and vegetables would grow on the upper floors while lower floors would house chickens and fish subsisting on the plant waste. Although Despommier's skyscraper farm has not yet been built, it popularized the idea of vertical farming and inspired many later designs.

Implementations of Vertical Farming

Developers and local governments in multiple cities have expressed interest in establishing a vertical farm: Incheon (South Korea), Abu Dhabi (United Arab Emirates), Dongtan (China), New York City, Portland, Los Angeles, Las Vegas, Seattle, Surrey, Toronto, Paris, Bangalore, Dubai, Shanghai, and Beijing.

In 2009, the world's first pilot production system was installed at Paignton Zoo Environmental Park in the United Kingdom. The project showcased vertical farming and provided a solid base to research sustainable urban food production. The produce is used to feed the zoo's animals while the project enables evaluation of the systems and provides an educational resource to advocate for change in unsustainable land-use practices that impact upon global biodiversity and ecosystem services.

In 2010 the Green Zionist Alliance proposed a resolution at the 36th World Zionist Congress calling on Keren Kayemet L'Yisrael to develop vertical farms in Israel. Moreover, a company named "Podponics" built a vertical farm in Atlanta consisting of over 100 stacked "growpods" in 2010 but reportedly went bankrupt in May 2016.

In 2012 the world's first commercial vertical farm was opened in Singapore, developed by Sky Greens Farms, and is three stories high. They currently have over 100 nine meter-tall towers.

In 2012, a company named The Plant debuted its newly-developed vertical farming system housed in an abandoned meatpacking building in Chicago, Illinois. The utilization of abandoned buildings to house vertical farms and other sustainable farming methods are a fact of the rapid urbanization of modern communities.

In 2013 the Association for Vertical Farming (AVF) was founded in Munich. By May 2015, the AVF had expanded with regional chapters all over Europe, Asia, USA, Canada and the United Kingdom. This organization unites growers and inventors to improve food security and sustainable development. The AVF focuses on advancing vertical farming technologies, designs and businesses by hosting international info-days, workshops, and summits.

In 2015 the London company, Growing Underground, began the production of leafy green produce underground in abandoned underground World War II tunnels.

In 2016, a startup called Local Roots launched the "TerraFarm", a vertical farming systems hosted in a 40-foot shipping container, which includes computer vision integrated with an artificial neural network to monitor the plants; and is remotely monitored from California. It is claimed that the TerraFarm system "has achieved cost parity with traditional, outdoor farming" with each unit producing the equivalent of "three to five acres of farmland," using 97% less water through water recapture and harvesting the evaporated water through the air conditioning. The first vertical farm in a US grocery store opened in Dallas, Texas in 2016.

In 2017, a Japanese company, Mirai, began marketing its multi-level vertical farming system with impressive statistics. The company states that it can produce 10,000 heads of lettuce a day - 100 times the amount that could be produced with traditional agricultural methods, because their special purpose LED lights can decrease growing times by a factor of 2.5. Additionally, this can all be achieved with 40% less energy usage, 80% less food waste, and 99% less water usage than in traditional farming methods. Further requests have been made to implement this technology in several other Asian countries.

In 2019, Kroger partnered with German startup Infarm to install modular vertical farms in two Seattle-area grocery stores.

Advantages

Agricultural Efficiency

Traditional farming's arable land requirements are too large and invasive to remain sustainable for future generations. With the ever-so-rapid population growth rates, it is expected that arable land per person will drop about 66% in 2050 in comparison to 1970. Vertical farming allows for, in some cases, over ten times the crop yield per acre than traditional methods. Unlike traditional farming in non-tropical areas, indoor farming can produce crops year-round. All-season farming multiplies the productivity of the farmed surface by a factor of 4 to 6 depending on the crop. With crops such as strawberries, the factor may be as high as 30.

Vertical farming also allows for the production of a larger variety of harvestable crops because of its usage of isolated crop sectors. As opposed to a traditional farm where one type of crop is harvested per season, vertical farms allow for a multitude of different crops to be grown and harvested at once due to their individual land plots.

Resistance to Weather

Crops grown in traditional outdoor farming depend on supportive weather and suffer from undesirable temperatures rain, monsoon, hailstorm, tornado, flooding, wildfires, and drought. "Three recent floods cost the United States billions of dollars in lost crops, with even more devastating losses in topsoil.

The issue of adverse weather conditions is especially relevant for arctic and sub-arctic areas like Alaska and northern Canada where traditional farming is largely impossible. Food insecurity has been a long-standing problem in remote northern communities where fresh produce has to be shipped large distances resulting in high costs and poor nutrition. Container-based farms can provide fresh produce year-round at a lower cost than shipping in supplies from more southerly locations with a number of farms operating in locations such as Churchill, Manitoba,and Unalaska, Alaska. As with disruption to crop growing, local container-based farms are also less susceptible to disruption than the long supply chains necessary to deliver traditionally grown produce to remote communities. Food prices in Churchill spiked substantially after floods in May and June 2017 forced the closure of the rail line that forms the only permanent overland connection between Churchill and the rest of Canada.

Environmental Conservation

Up to 20 units of outdoor farmland per unit of vertical farming could return to its natural state, due to vertical farming's increased productivity. Vertical farming would reduce the amount of farmland, thus saving many natural resources.

Deforestation and desertification caused by agricultural encroachment on natural biomes could be avoided. Producing food indoors reduces or eliminates conventional plowing, planting, and harvesting by farm machinery, protecting soil, and reducing emissions.

Traditional farming is often invasive to the native flora and fauna because it requires such a large area of arable land. One study showed that wood mouse populations dropped from 25 per hectare to 5 per hectare after harvest, estimating 10 animals killed per hectare each year with conventional farming. In comparison, vertical farming would cause nominal harm to wildlife because of its limited space usage.

Problems

Economics

Vertical farms must overcome the financial challenge of large startup costs. The initial building costs would exceed $100 million, for a 60 hectare vertical farm. Office occupancy costs can be high in major cities, with office space in cities such as Tokyo, Moscow, Mumbai, Dubai, Milan, Zurich, and Sao Paulo ranging from $1850 to $880 per square meter. Since vertical farms are meant to be located in the centers of major cities, owners of vertical farms would have to pay the occupancy costs that any other office in the same zone would have to pay. In Victoria, Australia a hypothetical 10-level vertical farm would have a cost of US$349 per square meter of arable land, whereas a traditional farm in rural Victoria would have a cost of US$0.40 per square meter of arable land. Eventually as vertical farms become more efficient, the startup cost will be overcome, however even a vertical farm with a yield per hectare factor 50 times larger than a traditional farm's yield, it would take 6–7 years for the vertical farm to break even in costs.

Opponents question the potential profitability of vertical farming. In order for vertical farms to be successful financially, high value crops must be grown since traditional farms provide low value crops like wheat at cheaper costs than a vertical farm. Louis Albright, a professor in biological and environmental engineering at Cornell stated that a loaf of bread that was made from wheat grown in a vertical farm would cost US$27. However, according to the US Bureau of Labor Statistics, the average loaf of bread cost US$1.296 in September 2019, clearly showing how crops grown in vertical farms will be non-competitive compared to crops grown in traditional outdoor farms. In order for vertical farms to be profitable, the costs of operating these farms must decrease.

The developers of the TerraFarm system produced from second hand, 40 foot shipping containers claimed that their system "has achieved cost parity with traditional, outdoor farming".

Energy Use

During the growing season, the sun shines on a vertical surface at an extreme angle such that much less light is available to crops than when they are planted on flat land. Therefore, supplemental light would be required. Bruce Bugbee claimed that the power demands of vertical farming would be uncompetitive with traditional farms using only natural light. Environmental writer George Monbiot calculated that the cost of providing enough supplementary light to grow the grain for a single loaf would be about $15. An article in the Economist argued that "even though crops growing in a glass skyscraper will get some natural sunlight during the day, it won't be enough" and "the cost of powering artificial lights will make indoor farming prohibitively expensive". Moreover, researchers determined that if only solar panels were to be used to meet the energy consumption of a vertical farm, "the area of solar panels required would need to be a factor of twenty times greater than the arable area on a multi-level indoor farm", which will be hard to accomplish with larger vertical farms. A hydroponic farm growing lettuce in Arizona would require 15,000 kJ of energy per kilogram of lettuce produced. To put this amount of energy into perspective, a traditional outdoor lettuce farm in Arizona only requires 1100 kJ of energy per kilogram of lettuce grown.

As the book by Dr. Dickson Despommier "The Vertical Farm" proposes a controlled environment, heating and cooling costs will resemble those of any other multiple story building. Plumbing and elevator systems are necessary to distribute nutrients and water. In the northern continental United States, fossil fuel heating cost can be over $200,000 per hectare. Research conducted in 2015 compared the growth of lettuce in Arizona using conventional agricultural methods and a hydroponic farm. They determined that heating and cooling made up more than 80% of the energy consumption in the hydroponic farm, with the heating and cooling needing 7400 kJ per kilogram of lettuce produced. According to the same study, the total energy consumption of the hydroponic farm is 90,000 kJ per kilogram of lettuce. If the energy consumption is not addressed, vertical farms may be an unsustainable alternative to traditional agriculture.

Pollution

If power needs are met by fossil fuels, the environmental effect may be a net loss; even building low-carbon capacity to power the farms may not make as much sense as simply leaving traditional farms in place, while burning less coal. Louis Albright argued that in a "closed-system urban farming based on electrically generated photosynthetic light", a pound of lettuce would result in 8 pounds of carbon dioxide being produced at a power plant, and 4,000 pounds of lettuce produced would be equivalent to the annual emissions of a family car. He also argues that the carbon footprint of tomatoes grown in a similar system would be twice as big as the carbon footprint of lettuce. However, lettuce produced in a greenhouse that allows for sunlight to reach the crops saw a 300 percent reduction in carbon dioxide emissions per head of lettuce. As vertical farm system become more efficient in harnessing sunlight, the less pollution they will produce.

Greenhouses commonly supplement carbon dioxide levels to 3–4 times the atmospheric rate. This increase in CO_2 increases photosynthesis rates by 50%, contributing to higher yields. Some greenhouses burn fossil fuels purely for this purpose, as other CO_2 sources, such as those from furnaces, contain pollutants such as sulphur dioxide and ethylene which significantly damage plants. This means a vertical farm requires a CO_2 source, most likely from combustion. Also, necessary ventilation may allow CO_2 to leak into the atmosphere.

Greenhouse growers commonly exploit photoperiodism in plants to control whether the plants are in a vegetative or reproductive stage. As part of this control, the lights stay on past sunset and before sunrise or periodically throughout the night. Single story greenhouses have attracted criticism over light pollution.

Hydroponic greenhouses regularly change the water, producing water containing fertilizers and pesticides that must be disposed of. The most common method of spreading the effluent over neighbouring farmland or wetlands would be more difficult for an urban vertical farm.

Building-integrated Agriculture

Building-integrated agriculture (BIA) is the practice of locating high performance hydroponic greenhouse farming systems on and in mixed-use buildings to exploit synergies between the built environment and agriculture.

Typical characteristics of BIA installations include: recirculating hydroponics, waste heat captured from a building's heating-ventilation-air condition system (HVAC), solar photovoltaics or other forms of renewable energy, rainwater catchment systems, and evaporative cooling.

The term building-integrated agriculture was coined by Ted Caplow in a paper delivered at the 2007 Passive and Low Energy Cooling Conference in Crete, Greece.

Applications of BIA are motivated by trends in patterns of energy use, global population, and global climate change. Specific observations include:

- Globally, modern agriculture uses 70% of fresh water withdrawals, is the world's largest source of water pollution, and the largest consumer of land.
- According to official UN estimates, global population is expected to exceed nine billion by 2050. Food travels hundreds of thousands of miles to reach urban consumers, adding to traffic congestion, air pollution and carbon emissions.
- Increased urbanization, results in marginalization of natural world and distance from food production.
- In the United States, buildings account for 39% of energy use, 68% of electricity consumption, 12% of water consumption, and 38% of carbon dioxide emissions.
- Global warming is predicted to lead to widespread shortages of food, water, and arable land by 2050.

Environmental Advantages

Proponents maintain that BIA is an environmentally sustainable strategy for urban food production that reduces our environmental footprint, cuts transportation costs, enhances food security / safety, conserves water, protects rivers, improves health, reduces waste, cools buildings, and combats global warming. For example, hydroponics uses ten to twenty times less land and ten times less water than conventional agriculture, while eliminating chemical pesticides, fertilizer runoff, and carbon emissions from farm machinery and long distance transport. Using a building's waste heat and solar photo voltaic panels reduces fossil fuel emissions that typically result from production and distribution. Rainwater catchment systems help to manage stormwater, much like a green roof.

Economics

Integrating a farm into a building offers all of the building performance benefits of a more conventional green roof, and results in a lower combined energy bill than if the components were separate. These systems are achievable with extant technology. Projects such as Gotham Greens 10,000-square-foot (930 m^2) greenhouse will cost approximately $1.4 million to build. Shulman, Robin. “Raising the Root: Some City Dwellers Are Hoping Rooftop Farming Will Bear Fruit”.

Applications

BIA systems may be integrated into commercial, educational, and residential buildings of varying sizes. Feasibility varies based on building size, climate, availability of light, and new build vs. retrofit. BIA farms are located on the building’s envelope to make maximum use of normal light. Both horizontal (rooftop) and vertical (façade) surfaces may be used.

A sprinkling of rooftop hydroponic greenhouses can be found around the world, including at academic centers in the United States; on a hospital in Singapore; in the Netherlands, in India, and in parts of the developing world. The Science Barge, while not on a building, is widely credited with invigorating interest in BIA in New York City, following its 2007 public debut.

Retrofit

Example retrofit projects include: Eli Zabar’s Vinegar Factory Greenhouse, which has been growing vegetables since 1995 heats his rooftop greenhouse with waste heat from the store’s bakery, and Gotham Greens, a company building New York City’s first commercial scale, hydroponic rooftop farm.

New Build

Example new build projects include the Forest Houses greenhouse, a fully integrated rooftop farm integrated onto the rooftop of an affordable housing complex, and Solar Two, an environmental learning center that will feature a Vertically Integrated Greenhouse.

Proposed

Proposed projects include Dongtan, a proposed eco-city island in Shanghai, and Masdar City, a carbon neutral city being built in Abu Dhabi, UAE and India.

Related Concepts

Vertical farming is a proposed agricultural concept in which entire urban high-rise buildings, not just the building envelope, are dedicated to large-scale farming. According to various researchers, to be realized vertical farms would require significant technological breakthroughs with regards to energy consumption and lighting. It has been estimated that a prototype five-story farm would cost between $20 million to $30 million.

In compost heated greenhouses, heat and carbon dioxide are generated from a manure-based compost contained in a special chamber attached to one side of the greenhouse. The New Alchemy Institute designed and built an experimental composting greenhouse in 1983 to research opportunities for the production of biothermal energy. Growing Power utilizes heat produced through vermicomposting to provide heat for their greenhouse.

Nutrient Film Technique

Nutrient film technique (NFT) is a hydroponic technique where in a very shallow stream of water containing all the dissolved nutrients required for plant growth is re-circulated past the bare roots of plants in a watertight gully, also known as channels.

Plants placed into nutrient-rich water channels in an NFT system.

A home-built NFT hydroponic system.

NFT was pioneered in 1965 by Allen Cooper at the Glasshouse Crops Research institute in Littlehampton, England. In an ideal system, the depth of the recirculating stream is very shallow, little more than a film of water, hence the name 'nutrient film'. This ensures that the thick root mat which develops in the bottom of the channel receives adequate air exposure.

A properly designed NFT system is based on using the right channel slope, the right flow rate, and the right channel length. The plant roots are exposed to adequate supplies of water, oxygen and nutrients. In earlier production systems, there was a conflict between the supply of these requirements, since excessive or deficient amounts of one results in an imbalance of one or both of the others.

NFT, because of its design, provides a system wherein all three requirements for healthy plant growth can be met at the same time, provided that the simple concept of NFT is always remembered and practiced. The result of these advantages is that higher yields of high-quality produce are obtained over an extended period of cropping. A downside of NFT is that it has very little buffering against interruptions in the flow, such as a result of a power outage. But, overall, it is one of the more productive techniques.

The same design characteristics apply to all conventional NFT systems. While slopes along channels of 1:100 have been recommended, in practice it is difficult to build a base for channels that is sufficiently true to enable nutrient films to flow without ponding in locally depressed areas. As a consequence, it is recommended that slopes of 1:30 to 1:40 be used. This allows for minor irregularities in the surface, but, even with these slopes, ponding and waterlogging may occur. The slope may be provided by the floor, or benches or racks may hold the channels and provide the required slope. Both methods are used and depend on local requirements, often determined by the site and crop requirements.

Flow Rates

As a general guide, flow rates for each gully should be 1 litre per minute. At planting, rates may be half this, and the upper limit of 2L/min appears about the maximum. Flow rates beyond these extremes are often associated with nutritional problems. Depressed growth rates of many crops have been observed when channels exceed 12 metres in length. On rapidly growing crops, tests have indicated that, while oxygen levels remain adequate, nitrogen may be depleted over the length of the gully. As a consequence, channel length should not exceed 10–15 metres. In situations where this is not possible, the reductions in growth can be eliminated by placing another nutrient feed halfway along the gully and reducing flow rates to 1L/min through each outlet. Care needs to be taken to maintain hygienic conditions and to avoid heavy metal contamination of NFT systems by using mainly plastic or stainless steel pumps and components.

Potato Minitubers

Most potato varieties are maintained in plant tissue culture and micropropagation methods are used to increase the amount of planting material. Since tissue culture plants perform poorly when planted into field soil, they are instead planted into greenhouses or screenhouses to generate tubers, which are referred to as minitubers. In many countries, it is common for NFT or aeroponic systems to be used for production of minitubers from tissue culture plantlets. The minitubers are planted into the field 6 to 14 months after harvest to grow a crop of potatoes. This first crop of field-grown potatoes is typically replanted to generate more potatoes rather than consumed.

Controversy

A leading proponent of NFT was Dr. Allen Cooper, a scientist at the Glasshouse Crops Research Station in England who published the book The ABC of NFT. NFT systems were used by a significant proportion of commercial growers in the UK through the 1980-1990 period but were only used for lettuce in Europe. Dutch growers particularly rejected NFT because of the perceived high risk of disease spread by the recirculating solution. NFT ensures that plants have unlimited access to water at all times, but it is now recognized that fruiting crops can benefit from carefully limited water supplies. Leafy crops like lettuce benefit from unlimited water supplies and are still widely grown using NFT, but now most commercial greenhouse crops of tomatoes, capsicums and cucumbers are grown hydroponically using some kind of inert media, with rockwool being the most important medium worldwide. NFT remains a very popular system for home use.

Ebb and Flow Sub-irrigation

Ebb and Flow (also called Ebb flood and flood drain) are two phases of the tide or any similar movement of water. The Ebb is the outgoing phase, when the tide drains away from the shore; and the flow is the incoming phase when water rises again. The terms are also common in figurative use.

Ebb and Flow is a form of hydroponics that is known for its simplicity, reliability of operation and low initial investment cost. Pots are filled with an inert medium which does not function like soil or contribute nutrition to the plants but which anchors the roots and functions as a temporary reserve of water and solvent mineral nutrients. The hydroponic solution alternately floods the system and is allowed to ebb away.

Under this system a water-tight growing bed, pots that are placed in the table will hold the medium. The medium can consist of material that the root system can reach the level ebb reaches. A medium can consist of lava rock, rockwool cubes, and fiber. A mixture of these different types of medium can absorb the moister more evenly and efficiently. this is periodically flooded for a short period (5 to 10 minutes) with a nutrient solution pumped from a supply tank. By placing the nutrient solution supply tank below the growing bed, the nutrient solution can drain back by gravity. This hydroponic growing system can be very effective for any size plants depending on space. The method is inefficient in its use of water and plant nutrient reagents. Root disease occurrence and nutrient element insufficiency can occur without repeated use of the solution without proper oxygenation on the tables that are flooded. Because it is a "closed" system, the re-circulated nutrient solution will require monitoring every time the system flows back the reservoir for pH and PPM. Within the growing period, the nutrient solution may require replacement. The rooting medium will require washing to remove root debris and accumulated precipitates as well as sterilization before reuse.

Principles of Operation

The fundamental principle of hydroponics relies on fertilized and aerated water which provides both nutrition and oxygen to a plant's root zone. It often involves relatively sophisticated mechanization processes which can be daunting to casual hobbyists. Nutrient solutions must usually be below the temperature at which pathogen growth can begin, yet not so cool that root activity is suppressed. Active aeration of the fertilizer solution is common, since root systems themselves remove oxygen, creating conditions which also can promote pathogenic bacteria and water-borne molds.

E&F utilizes the fact that the solution is not left in constant contact with the roots of plants, to avoid the need for oxygenating or chilling of the solution. Instead it relies on characteristics of root function to provide passive oxygenation at a high level which tends to suppress pathogen growth.

Simplicity is maintained through usage of a single, two-directional path for the solution. Water flows in and out using the same tube. When the pump has raised water into the tray, briefly submerging the roots, the pump is rendered inactive using a switch, typically a timer, and the water flows back down the same tube. This eliminates the need for more than one sealed fitting and reduces overall complexity of the system.

Ebb and flow systems come on according to the water-holding capacity of the medium in which the roots sit. Highly water-retentive media can require watering only once a day, while others require two to as many as six floodings, with each "flood" stage only lasting a few minutes. The time it takes to flood the roots is not a critical parameter, which means that pumps are often moderate in capacity and can be small for systems sustaining indoor plants. This makes the method popular with amateur and urban gardeners. Gravity acts as drain pump, and aeration is accomplished through thin-filming and positive displacement of air as it is forced out of the root zone by water.

Aeration in Ebb and Flood Systems

Aeration of an ebb and flood system is an important aspect of its operation. Automatic displacement eliminates air which has been de-oxygenated by the roots as the water rises to its highest flood stage. When the pump turns off, gravity pulls the water downwards, which re-exposes the space around the roots to the air.

The film of water left around the roots during ebb has a high surface-to-mass ratio, which means that even as the roots absorb oxygen, its high surface area facilitates re-oxygenation, which can sustain the roots as long as their surfaces remain damp. The high oxygen content of water filmed in this way suppresses most harmful lifeforms, keeping the root zones disease free. In other types of hydroponics this function must be performed by cooling the solution to protect it from pythium, a form of water mold responsible for a condition called 'root rot', in which the outer cells of the roots die, turn brown and slough off when handled. Need for supplementary oxygenation using air pumps is also eliminated, which increases reliability and reduces complexity.

Ebb and flow hydroponic systems are also quiet, while using less power than other hydroponic systems, which means that they can be used in environments where acoustic signature and excessive plumbing is objectionable, such as residential or classroom applications where space is at a premium.

Drawbacks to E&F Systems

Ebb and flow systems are flexible, with few practical drawbacks. Though typically known for compact cultivation of plants having smaller stature, it has been used for growing large plants, using buckets ranging in size from 1 gallon to 5 gallons, making use of high-volume pumps such as those in large aquariums, decorative fountains and koi ponds.

There are facets to these systems that present some labor investment in large-scale applications. These are primarily management of media between uses, such as washing and sterilization. This can be done by dumping into the tray and filling with a sterilizing solution such as hydrogen peroxide or chlorine solution, temporarily plugging the drain, with hand removal of root fragments. Larger containers require transferring the media to a suitable surface after sterilization to permit removal of leftover plant material.

A second drawback is that the roots tend to grow together, meaning removal of harvested or damaged plants can be somewhat problematic in plants having well-developed root systems. Commercial crops harvested at one time are somewhat immune to concerns related to that aspect of the system, but in the event of pathogenic invasion the problem can quickly spread, as all the roots share the same flood source.

Also, most ebb and flow systems use a recycling reservoir to flood the table. Over a period of time the pH of the nutrient solution may fluctuate to a range which is unhealthy for the plant. If the pH is not corrected, various problems may occur, including but not limited to poor nutrient absorption and leaf cannibalization. As the name implies, leaf cannibalization occurs as the plant takes nutrients from one part of the plant and uses those nutrients in a different part of the plant. Leaf cannibalization appears as yellow or brown spots on leaves.

During Flower pH rises quite a bit each day. It is best to adjust first thing and last thing each day. Also, during Flower nutrients and water absorption increases, while root exudate gets carried back to the reservoir. This causes ppms to increase significantly. Proper control requires routine checking and replacing with fresh nutrients ~ 5 days to avoid toxicity.

Since the plants is being fed several times a day, lower ppm nutrients (600–800) are sufficient. Pushing with higher ppms can cause the plant to burn up from the inside, especially when significant water evaporation/usage causes the remaining nutrient concentration to increase beyond 1500 ppms.

Choosing a medium like lava rock is ideal for flood and drain in that it drains quickly, and due to its rough texture, it traps small amounts of oxygen and nutrients which keeps the root zone moist between feedings. Also, because it drains quickly, the number of feedings can be increased to roughly every 45–60 minutes during lights on, producing explosive growth. Should the grower opt for this method, the nutrient ppm should be kept below 800, better still 600.

Poor drainage, or incomplete drainage, may cause a condition wherein dense roots are exposed to stagnant water which is trapped by the root mass. Root rot and fungal growth are the most common result of stagnant water. Some E&F systems are not as immune to root rot as a well-designed system would be. In tables where plants are larger than optimal for the system, this can create the need for modifications such as screens or beds of medium-sized gravel to prevent standing water. Tilting the tray is one way to achieve better drain characteristics. In bucket E&F this problem can be dealt with in a similar manner, ensuring good drainage through using medium of adequate size and ensuring that drainage of the container between flood cycles is complete.

Hydrogen peroxide is also added to nutrient solutions when there is suspicion that the anaerobic pythium root rot mold has begun to proliferate on the roots' surfaces. The oxygen liberated from the hydrogen peroxide is destructive to single-celled organisms and is administered in dosages which vary with the concentration of the peroxide. Typically several tablespoons or more of 3.5% peroxide solution per gallon of water are used. The temporary rise in the oxygen level is only minimally damaging to roots, while eradicating the water-borne mold can significantly increase yield or even save a crop's viability.

Importance of Ebb and Flow Method

It is important to irrigate your plants whether you are growing them for sale or home use. For uniform growth of plants, they are supposed to be irrigated and nurtured properly. There are various methodologies used in growing and irrigating plants a gardener can choose from. One of the most effective and simple to manage is the hydroponic ebb and flow or flood and drain system.This form of sub irrigation offers lot of benefits to the gardener in terms of simplicity, cost effectiveness and maintenance. The setup and principle of operation of this system is very simple and can easily be mastered. The mechanism works by temporarily flooding a grow tray with water full of nutrients or a nutrient solution. After some specific timing, the nutrient solution is then drained back to a reservoir. The cycle is repeated under the control of a timer ensuring the plants have access to nutrients and water whenever they need it.The frequency of the flooding and draining the water containing the nutrients by the timer is dependent on various factors. The type of plant being grown determines how many times in a day will the grow tray be flooded since different plants need different amounts of water and nutrients for growing optimally.The type of growing media, depending on its rate of holding water can also determine the rate of flooding.

The growing materials commonly used are gravel, granular grow rocks and Rock wool. Other factors that can be considered are the prevailing day temperatures and the humidity levels.Some of the advantages making this mechanism popular to serous gardeners other than other methods of irrigation include the fact that plants get abundant nutrients. By timing the flow of nutrients, plants are able to obtain maximum nutrients and hence grow well. It is even easier to manage the system since it is not labor intensive and does not require extensive knowledge in hydroponics to operate it.It is also easy to construct and assemble a hydroponic system. The initial cost of materials can be daunting but if you consider the long term benefits of these project you will realize its benefits. The size of the garden and the number of plants to be grown should be considered first before construction.Hydroponic ebb and flow or flood and drain systems however have some setbacks too.

The principle of operation depends on power supply for pumping water to the grow tray and draining the same to the reservoir. Whenever power outbreaks or a mechanical fault occur, it can affect the cycle of pumping and draining nutrients. This can adversely affect the plants being grown especially those that require constant supply of water and nutrients. This problem can however be temporarily handled by using a growing media that retains more water.The draining of water from the grow tray to the reservoir can alter the pH levels of the solution. Changes in pH can have adverse effects on the plants and if unchecked it can cause toxicity to the plants. Continuous draining of water over time may also initiate salt build up on the plant roots or the growth media and this can block vital chemical components from reaching the plant leading to deficiencies. Proper maintenance and frequent checks on the nutrient solutions will solve these setbacks.

Disadvantages of using an Ebb and Flow Hydroponics System

- Higher Risk: There are several risks involved with using an ebb and flow system: the flood/drain component could malfunction, causing the flood of the area you've installed the ebb and flow in. Another risk is that of ruining your plant yields, due to the flooding mechanism. Because many plants require a lot of water on their roots, but need not get their corona wet, ebb and flow hydroponics systems can actually be dangerous to your plants. The strawberry is the most well known plant that can suffer from this in an ebb and flow hydroponics system.

- Breakdowns: Because the ebb and flow system is ran with the help of a pump controller, which is usually prone to breaking down, the risk of the entire system breaking down is also higher than with many other system types.
- Plant Toxicity/Deficiency: Ebb and flow hydroponics systems flood the plants with water, allow them to suck up their required nutrients, then drain the water back in the tank. Over time, continuously draining the water will have salt building up on the roots or the growth media, which will block out certain chemical elements from reaching the plant, causing a deficiency.
- Unstable pH levels: When water and nutrient solutions are drained back from the growth tray and into the nutrient container, the pH level of the solution changes dramatically. This can have devastating effects on your plants in time, blocking out certain chemical elements, or allowing others to run out of control and cause toxicities.

Sand and Gravel

Sand and gravel are both potentially very suitable growing media for hydroponic systems. The above is 10mm "sharp" gravel, which is commonly used in aquarium filtration systems, and could be used more often for hydroponics. Note the irregular sizes/shapes, and the sharp edges, as opposed to pea gravel which is much more rounded.

Advantages and Disadvantages of Gravel

One extremely common growing medium is plain old gravel. While gravel comes in a range of sizes and shapes, from sharp to polished, the rule of thumb for hydroponics is to use 3/8 to 5/8 minus, which means particle size will pass through a 3/8" or 5/8" screen. So that size is the maximum particle size. Additionally, most gravel users will wash their gravel before initial use, since some gravel comes with a high percentage of "fines", ie dust and dirt which would only serve to clog up filters. Gravel often is also designated as either "sharp" or "pea" gravel, to differentiate between jagged edges and rounded/weathered edges. Very generally speaking, sharp gravel is preferred because all those edges provide more surface area. Surface area is important because that's what holds minute quantities of moisture in between irrigation cycles. The sharp edges also help keep the gravel mass from shifting, and gives the roots a better anchor than smooth polished pebbles would provide. However, pea gravel will work if that's the only type available. You might need to cycle your nutrient solution more frequently to compensate.

Other advantages include gravel's relatively low cost and easy availability. Furthermore, it can be reused indefinitely. All the grower needs to do after each crop is to let the material dry out, then spread out in a large area, wash down to remove dead roots and other debris, rinse thoroughly, then put back into the grow beds. Few growing media offer such longevity particularly at such a low price. Yet gravel has its drawbacks. First, it is amazingly heavy. A small 2 quart bag of gravel weighs nearly 50 pounds. Any planting beds using gravel must be built to hold all that weight, particularly if they are to be raised above ground level for easy access. Secondly, gravel can change the pH of the water moving through the system. Both hydroponics and aquaponics rely on fairly narrow pH ranges for optimum nutrient availability. Gravel can change that pH depending on the parent rock material. Some types of rock will chemically react with the water enough to raise the pH levels far higher than they should be. Sometimes, that swing cannot be corrected even with acidifiers. If your water source is also alkaline, that single problem can virtually guarantee your pH would never be where it should be. Or it will cost a lot to fix it. On the other hand, if your water source happens to be slightly acidic, that gravel can provide built-in correction. So it pays to know your water's natural pH and then work from that basis as you select your growing media. And finally, gravel has no water holding ability. If you are using drip irrigation that may not be a problem. But if you want to use flood and drain, you may want to provide for backup irrigation or power supplies to make sure your plants are never left high and dry in case of a power outage, or plumbing blockage.

Advantages and Disadvantages of Sand

All the comments made about gravel also apply to sand, with one exception. The sand particles are already jagged enough that there is really no concern about sufficient surface area or anchoring characteristics. It is even heavier than gravel, so planting beds must be robust to support all that weight even if at ground level. It does not have any better water retention characteristics compared to gravel, but it does have better wicking abilities. So if a reservoir was built at the bottom of each planting bed, that water could wick up into the root zone more easily with sand growing media, as opposed to gravel growing media. That's not enough to supply your plants for long, but it can provide a buffer during power outages or other irrigation interruptions.

The major disadvantages to sand are the weight, and the extremely small void size between the sand particles. Many growers who have experimented with both sand and gravel eventually end up with gravel because sand is just a little too heavy and hard to irrigate. Another related disadvantage is that filters have to be smaller than usual to ensure the sand doesn't run out through the irrigation cycle drain. That means the filters have to be checked and cleared more often, usually by hand.

One instance where sand really shines is when folks have access to relatively large gravel types and need "filler" which can be rinsed repeatedly as part of a normal irrigation cycle. In that sense, sand and gravel can be a wonderful combination.

As we saw with gravel, sometimes sand becomes the growing medium of choice simply because it is cheap and widely available. Those characteristics can overcome a lot of other disadvantages in some circumstances. If you have access to cheap or free sand, it's certainly worth experimenting a bit to find out if it might work well for you.

References

- Static-hydroponic-systems-cons-and-pros: scienceinhydroponics.com, Retrieved 23 April, 2020
- NFT-continuous-flow-solution-culture: growland.biz, Retrieved 16 June, 2020
- Fogponics, definition-160: maximumyield.com, Retrieved 19 May, 2020
- Anthroponics-report: hemmaodlat.se, Retrieved 18 July, 2020
- What-is-aquaponics-and-how-does-it-work: permaculturenews.org, Retrieved 10 January, 2020
- Disadvantages-passive-subirrigation-95169: homeguides.sfgate.com, Retrieved 13 March, 2020
- Rotary-hydroponics-vegetables-countertop: intelligentliving.co, Retrieved 04 February, 2020
- How-to-build-a-raft-hydroponic-system: thehydroponicsplanet.com, Retrieved 12 June, 2020
- Autopot-systems, hydroponics-systems: hydroponics.name, Retrieved 18 August, 2020

3
Nutrient Solutions

The hydroponic nutrient solution is a liquid which consists of minerals in the raw water and nutrients added with fertilizers. Minerals such as calcium, magnesium, sulfur and trace elements such as boron, manganese, iron and zinc may be present in the source water. This chapter has been carefully written to provide an extensive understanding of nutrient solutions.

A nutrient solution for hydroponic systems is an aqueous solution containing mainly inorganics ions from soluble salts of essential elements for higher plants. Eventually, some organic compounds such as iron chelates may be present. An essential element has a clear physiological role and its absence prevents the complete plant life cycle. Currently 17 elements are considered essential for most plants, these are carbon, hydrogen, oxygen, nitrogen, phosphorus, potassium, calcium, magnesium, sulphur, iron, copper, zinc, manganese, molybdenum, boron, chlorine and nickel. With the exception of carbon (C) and oxygen (O), which are supplied from the atmosphere, the essential elements are obtained from the growth medium. Other elements such as sodium, silicon, vanadium, selenium, cobalt, aluminum and iodine among others, are considered beneficial because some of them can stimulate the growth, or can compensate the toxic effects of other elements, or may replace essential nutrients in a less specific role. The most basic nutrient solutions consider in its composition only nitrogen, phosphorus, potassium, calcium, magnesium and sulphur; and they are supplemented with micronutrients.

The nutrient composition determines electrical conductivity and osmotic potential of the solution. Moreover, there are other parameters that define a nutrient solution as discussed below in detail.

pH of the Nutrient Solution

The pH is a parameter that measures the acidity or alkalinity of a solution. This value indicates the relationship between the concentration of free ions H^+ and OH^- present in a solution and ranges between 0 and 14.

In soil, the Troug diagram illustrates the pH effect on the availability of nutrients to plants. Similarly, changing the pH of a nutrient solution affects its composition, elemental speciation and bioavailability. The term "speciation" indicates the distribution of elements among their various chemical and physical forms like: free ions, soluble complexes, chelates, ion pairs, solid and gaseous phases and different oxidation states.

An important feature of the nutrient solutions is that they must contain the ions in solution and in chemical forms that can be absorbed by plants, so in hydroponic systems the plant productivity is closely related with to nutrient uptake and the pH regulation. Each nutrient shows differential responses to changes in pH of the nutrient solution as described below.

In the nutrient solution, NH_3 only forms a complex with H^+. For a pH range between 2 and 7, NH_3 is completely present as NH_4^+. Increasing the pH above 7 the concentration of NH_4^+ decreases, while the concentration of NH_3 augments.

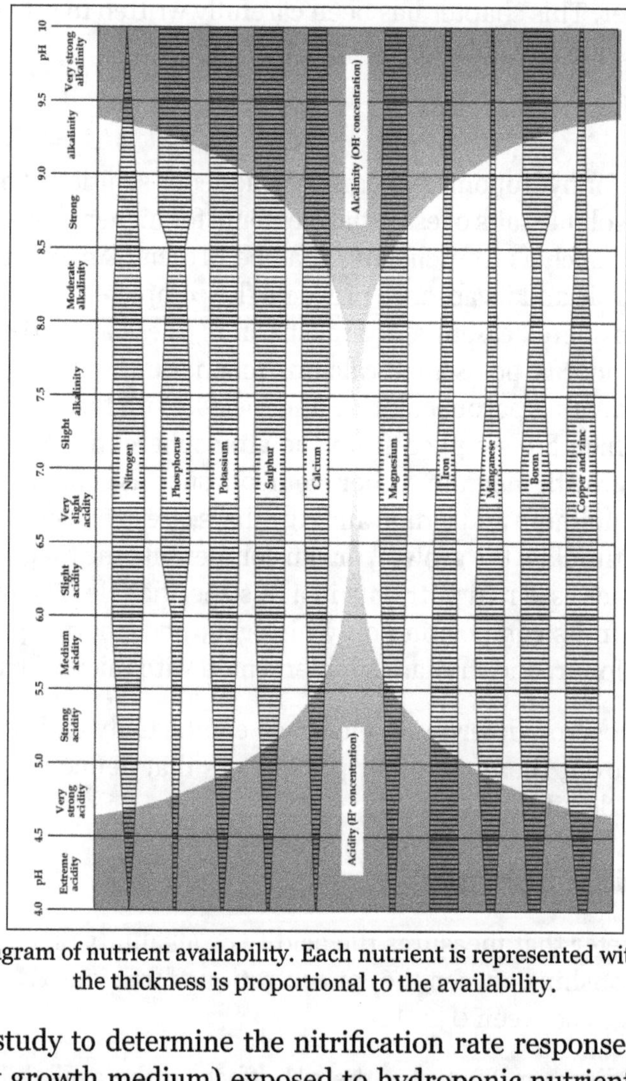

Troug diagram of nutrient availability. Each nutrient is represented with a band; the thickness is proportional to the availability.

Tyson et al. in a study to determine the nitrification rate response in a perlite trickling biofilter (root growth medium) exposed to hydroponic nutrient solution, varying

NO_3^- concentrations and two pH levels (6.5 and 8.5), founded that nitrification was significantly impacted by water pH. The increased ammonia oxidation rate (1.75) compared to nitrite oxidation rate (1.3) at pH 8.5 resulted in accumulation of NO_2^- to levels near those harmful to plants (observed peak of 4.2 mg L^{-1} NO_2^-. The potential for increased levels of un-ionized ammonia, which reduced plant nutrient uptake from micronutrient precipitation, are additional problems associated with pH 8.5.

Phosphorus is an element which occurs in forms that are strongly dependent on environment pH. In the root zone this element can be found as PO_4^{3-},HPO_1^{2-}, and $H_2PO_4^-$ ions; the last two ions are the main forms of P taken by plants. On inert substrates, the largest amount of P available in a nutrient solution is presented when its pH is slightly acidic (pH 5). In alkaline and highly acidic solutions the concentration of P decreases in a significant way. Namely, with pH 5, 100% of P is present as $H_2PO_4^-$; this form converts into HPO_4^{-2} at pH 7.3 (pKa2), reaching 100% at pH 10. The pH range that dominates the ion $H_2PO_4^{-2}$ on $H_2PO_4^-$ is between 5 and 6. The pH-dependent speciation of P is showed in figure.

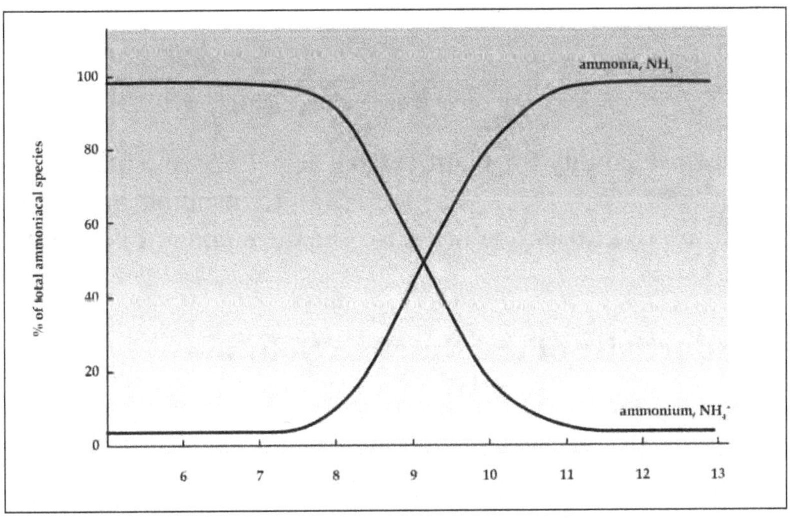

Ammoniacal speciation in function of pH.

Potassium is almost completely present as a free ion in a nutrient solution with pH values from 2 to 9; only small amounts of K^+ can form a soluble complex with SO_4^{2-} or can be bound to Cl^-. Like potassium, calcium and magnesium are available to plants in a wide range of pH; however, the presence of other ions interferes in their availability due to the formation of compounds with different grade of solubility. As water naturally contains HCO_3^-, this anion turns into CO_3^{-2} when the pH is higher than 8.3 or to H_2CO_3 when it is less than 3.5; the H_2CO_3 is in chemical equilibrium with the carbon dioxide in the atmosphere. Thus at a pH above 8.3, Ca^{2+} and Mg^{2+} ions easily precipitate as carbonates. Also, as mentioned above, when the pH of the nutrient solution increases, the HPO_4^{2-} ion predominates, which precipitates with Ca^{2+} when the product of the concentration of these ions is greater than 2.2, expressed in mol m-3. Sulphate also forms relatively strong complexes with Ca^{2+} and Mg^{2+}. As pH increases from 2 to 9, the amount of SO_4^{2-}, forming soluble complexes with Mg^{2+} as $MgSO_4$ and with K^+ as KSO_4^- increases.

Iron, copper, zinc, boron, and manganese, become unavailable at pH higher than 6.5. In Triticum aestivum, the manganese precipitation on root surfaces was correlated with a plant-induced rise in pH of culture above 5.5. Boron is mainly uptaken by plants as boric acid, which is not dissociated until pH is close to 7; to greater pH values, boric acid accepts hydroxide ions to form anionic species.

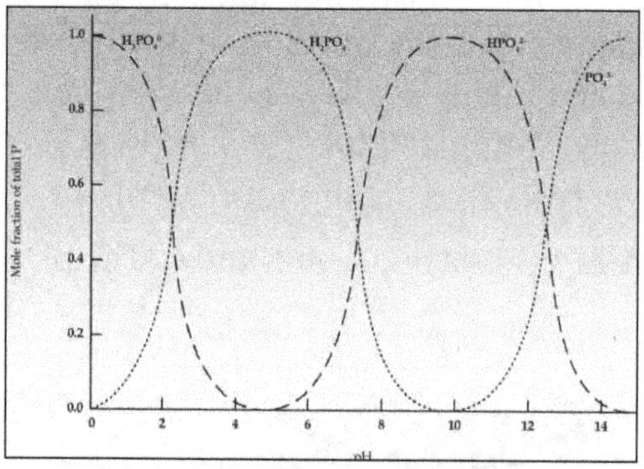

Speciation of P depending on pH.

Therefore, nutrient availability for plant uptake at pH above 7 may be restricted due to precipitation of Fe^{2+}, Mn^{2+}, PO_3^{4-}, Ca^{2+} and Mg^{2+} to insoluble and unavailable salts. The proper pH values of nutrient solution for the development of crops, lies between 5.5 and 6.5.

Electrical Conductivity of the Nutrient Solution

The total ionic concentration of a nutrient solution determines the growth, development and production of plants. The total amount of ions of dissolved salts in the nutrient solution exerts a force called osmotic pressure (OP), which is a colligative property of the nutrient solutions and it is clearly dependent of the amount of dissolved solutes. Also, the terms solute potential or osmotic potential are widely used in nutrient solution, which represent the effect of dissolved solutes on water potential; solutes reduce the free energy of water by diluting the water. Thus, the terms osmotic pressure and osmotic potential can be used interchangeably, still important considering the units that are used, commonly atm, bar and MPa.

An indirect way to estimate the osmotic pressure of the nutrient solution is the electrical conductivity (EC), an index of salt concentration that defines the total amount of salts in a solution. Hence, EC of the nutrient solution is a good indicator of the amount of available ions to the plants in the root zone. Estimation of the osmotic pressure of a nutrient solution from EC can be done by using the following empirical relations.

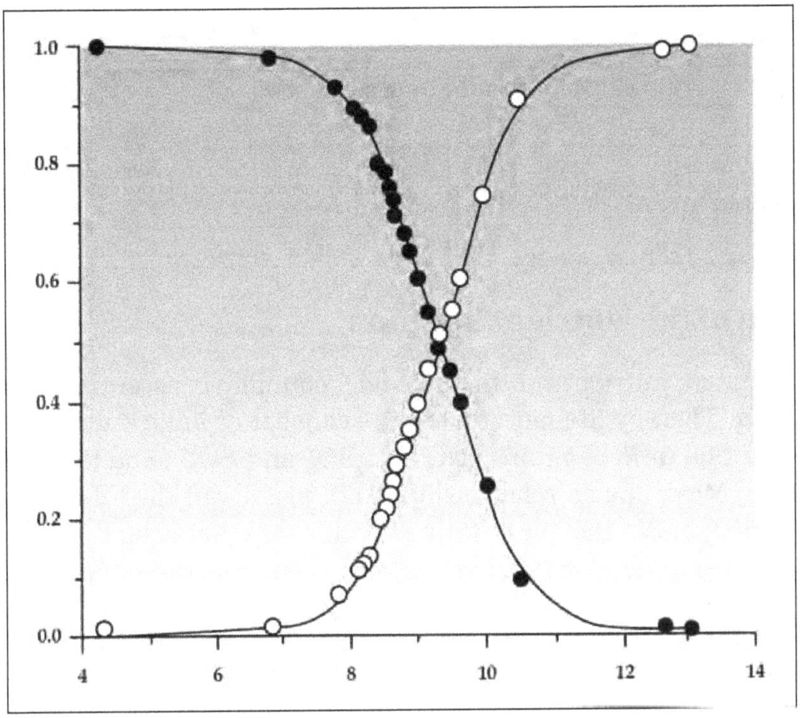

Transformation of boric acid (black circles) and anion forms of boric acid (white circles) as a function of Ph.

$$\text{OP}\ (\text{atm}) = 0.36 \text{ X EC } \left(\text{in dS m}^{-1} \text{ at } 25^{\circ}\text{C}\right)$$

$$\text{OP}\ (\text{bar}) = -0.36 \text{ X EC } \left(\text{in dS m}^{-1} \text{ at } 25^{\circ}\text{C}\right)$$

$$\text{OP}\ (\text{MPa}) = \text{OP}\ (\text{bars}) \text{ X } 0.1$$

The ions associated with EC are Ca^{2+}, Mg^{2+}, K^+, Na^+, H^+, NO_3^-, SO_4^{2-}, Cl^-, HCO_3^-, OH^-. The supply of micronutriments, namely Fe, Cu, Zn, Mn, B, Mo, and Ni, are very small in ratio to the others elements (macronutrients), so it has no a significant effect on EC.

The ideal EC is specific for each crop and dependent on environmental conditions; however, the EC values for hydroponic systems range from 1.5 to 2.5 ds m^{-1}. Higher EC hinders nutrient uptake by increasing osmotic pressure, whereas lower EC may severely affect plant health and yield. The decrease in water uptake is strongly and linearly correlated to EC. Some crops can grow with high levels of EC and even a proper management of EC of the nutrient solution can provide and effective tool to improve vegetable quality. In particular, parameters of fruit quality such as soluble solids content, titratable acidity and dry matter augmented by increasing EC level of nutrient solution from 2 to 10 dS m^{-1}. As a consequence, deep sea water (DSW) is being used for nutrient solution due to its high amount of Na^+, Mg^{2+}, K^+ and Ca^{2+}.

Table: Threshold EC for salinity groups and example of crops.

Salinity group	Threshold EC, dS m^{-1}	Example of crops
Sensitive	1.4	lettuce, carrot, strawberry, onion
Moderately sensitive	3.0	broccoli, cabbage, tomato, cucumber, radish, pepper
Moderately tolerant	6.0	soybean, ryegrass
Tolerant	10.0	bermuda-grass, sugarbeet, cotton

Composition of the Nutrient Solution

As previously stated, nutrient solutions usually contain six essential nutrients: N, P, S, K, Ca and Mg. Thereby Steiner created the concept of ionic mutual ratio which is based on the mutual ratio of anions: NO_3^-, $H_2PO_4^-$ and SO_4^{2-}, and the mutual ratio of cations K^+, Ca^{2+}, Mg^{2+}. Such a relationship is not just about the total amount of each ion in the solution, but in the quantitative relationship that keep the ions together; if improper relationship between them take place, plan performance can be negatively affected.

In this way, the ionic balance constraint makes it impossible to supply one ion without introducing a counter ion. A change in the concentration of one ion must be accompanied by either a corresponding change for an ion of the opposite charge, a complementary change for other ions of the same charge, or both.

When a nutrient solution is applied continuously, plants can uptake ions at very low concentrations. So, it has been reported than a high proportion of the nutrients are not used by plants or their uptake does not impact the production. For example, it was determined that in anthurium, 60% of nutrients are lost in the leachate; but in closed systems, however, the loss of nutrients from the root environment is brought to a minimum. Also it has been shown that the concentration of nutrient solution can be reduced by 50% without any adverse effect on biomass and quality in gerbera and geranium. Accordingly, Siddiqi et al. reported no adverse effect on growth, fruit yield and fruit quality in tomato when reduction of macronutrient concentrations to 50% of the control level as well as cessation of replenishment of the feed solution for 16 days after 7 months of growth at control levels were applied. However, it is expected that in particular situations, too low concentrations do not cover the minimum demand of certain nutrients.

On the other hand, high concentrated nutrient solutions lead to excessive nutrient uptake and therefore toxic effects may be expected. Conversely, there are evidences of positive effects of high concentrations of nutrient solution. In salvia, the increase of Hoagland concentration at 200% caused that plants flowered 8 days previous to the plants at low concentrations, increasing total dry weight and leaf area. Likewise, high levels of K^+ in the nutrient solution (14.2 meq L^{-1} vs 3.4 meq L^{-1}) increased fruit dry matter, total soluble solids content and lycopene concentration of tomato.

The explanation of these apparent controversial responses is the existence of optimal concentrations of certain nutrients in a solution for a culture under special environmental conditions as well as their relative proportions and not their absolute concentrations as determining factors. In order to prevent contradictory observations, Dufour & Guérin recommend: a) to monitor the availability of nutrients through changes in the ionic composition of the substrate by analysis of percolate, and b) to asses plant nutrient uptake by nutrient content analysis in leaves. Moreover, Voogt indicates that the nutrient solution composition must reflect the uptake ratios of individual elements by the crop and as the demand between species differs, the basic composition of a nutrient solution is specific for each crop. It must also be taken into account that the uptake differs between elements and the system used. For instance, in open-systems with free drainage, much of the nutrient solution is lost by leachate.

There are several formulations of nutrient solutions. Nevertheless, most of them are empirically based. Table comprises some of them.

Table: Concentration ranges of essential mineral elements according to various authors.

Nutrient	Hoagland & Arnon	Hewitt	Cooper	Steiner
	mg L^{-1}			
N	210	168	200-236	168
P	31	41	60	31
K	234	156	300	273
Ca	160	160	170-185	180
Mg	34	36	50	48
S	34	48	68	336
Fe	2.5	2.8	12	2-4
Cu	0.02	0.064	0.1	0.02
Zn	0.05	0.065	0.1	0.11
Mn	0.5	0.54	2.0	0.62
B	0.5	0.54	0.3	0.44
Mo	0.01	0.04	0.2	Not considered

Temperature of the Nutrient Solution

The temperature of the nutrient solution influences the uptake of water and nutrients differentially by the crop.

Two nutrient solution temperatures (cold and warm solution, 10 and 22 °C, respectively) were evaluated during two flowering events of rose plants (Rosa × hybrida cv. Grand Gala). Generally, cold solution increased NO_3^- uptake and thin-white roots production, but decreased water uptake. Nutrient solution temperature also had an effect on the photosynthetic apparatus. In general terms, the effective quantum yield and the fraction of open PSII reaction centres were higher in rose plants grown at cold solution.

In spinach seedlings, three temperatures of irrigation water (24, 26 and 28 °C) were evaluated during 8 weeks. Leaf length, leaf number and total fresh and dry biomass weights per plant were higher in plants grown at elevated temperatures, with optimum growth being recorded at 28 °C.

In tomato plants, rates of water and nutrients uptake by roots (which varied depending on solar radiation) were studied. High solution temperature (35 °C) produced effects in the short and long-term. In the short-term, water and nutrients uptake were activated through a decrease in water viscosity, and membrane transport was affected. In the long term, oxygen solubility was reduced, while enzymatic oxidization of phenolic compounds in root epidermal and cortex tissues were stimulated, but nutrient concentration in root xylem sap diminished, and the root xylem sap concentration of N, K, Ca became lower than those in the nutrient solution.

At temperatures below 22 °C the dissolved oxygen in the nutrient solution is sufficient to cover the demand of this element in tomato plants. Nevertheless, the requirement diminished as a consequence of a reduction in a number of physiological processes, including respiration, which further impacts plant growth. Conversely, temperatures over 22 °C, oxygen demand is not covered by the nutrient solution as higher temperatures increase the diffusion of this gas. At high temperatures of the nutrient solution an increased vegetative growth to a greater extent than desirable is observed, which reduces fructification.

To assess the importance of temperature on the solubility of oxygen, Table depicts data for temperatures that are usually filed within greenhouses, so that temperature has a direct relationship to the amount of oxygen consumed by the plant and reverse relationship with dissolved oxygen from the nutrient solution.

Temperature, °C	Oxygen solubility, mg L^{-1} of pure water
10	11.29
15	10.08
20	9.09
25	8.26
30	7.56
35	6.95
40	6.41
45	5.93

Nutrient Solution Management

Soilless cultivation allows a more accurate control of environmental conditions that offers possibilities for increasing production and improving quality of crops. In particular, in the nutrient solution parameters such as temperature, pH, electrical conductivity, oxygen content, among others can be manipulated. If these parameters are not controlled properly and in timing, advantage can be translated into disadvantages. Then, several ways to control some parameters of the nutrient solution are to be reviewed in the following:

pH Regulation

The pH value determines the nutrient availability for plants. Accordingly, its adjustment must be done daily due to the lower buffering capacity of soilless systems.

The changes in the pH of a nutrient solution depending on the difference in the magnitude of nutrient uptake by plants, in terms of the balance of anions over cations. When the anions are uptaken in higher concentrations than cations, for example nitrate, the plant excretes OH^- or HCO_3^- anions, to balance the electrical charges inside, which produces increasing in the pH value. This process is called physiological alkalinity.

Hence, incorporation of ammonium as N source in the nutrient solution regulates the pH and therefore nutrient availability is ensured. Breteler & Smit reported that ammonium depressed the pH of nutrient solution even in the presence of nitrate. In rose plants, the addition of ammonium in a nutrient solution containing nitrate produced a total nitrogen uptake increase during shoot elongation; and an increase in P concentration in the roots. The proportion of total nitrogen is added to the nutrient solution as ammonium is dependent on the crop.

On other hand, the chemical adjustment is widely used, namely the addition of acids to reduce the pH value. The pH is closely related to the concentration of HCO_3^- and CO_3^{2-} ; when an acid is applied, the CO_3^{2-} ion is transformed to HCO_3^-, and then HCO_3^- is converted into H_2CO_3. Carbonic acid is partially dissociated in H_2O and CO_2. Regulation of pH is normally carried out by using nitric, sulphuric or phosphoric acid, and such acids can be used either individually or combined.

Electrical Conductivity Management

Electrical conductivity (EC) is modified by plants as they absorb nutrients and water from the nutrient solution. Therefore, a decrease in the concentration of some ions is and an increase in the concentration of others is observed simultaneously, both in close and open systems. For example, in a closed hydroponic system with a rose crop, the composition of the nutrient solution in the tank was measured. It was observed that the concentration of Fe decreased very fast, while that of Ca^{2+}, Mg^{2+} and Cl^- increased; moreover, concentrations of K^+, Ca^{2+} and SO_4^{2-} did not reach critical levels. Instead, in an open system with recirculation of nutrient solution, an increase in the EC value due to the accumulation of high levels of some ions like bicarbonates, sulphates and chlorides is observed. So, the recycling of nutrient solution represents a point of discussion. Moreover, the substrates can retain ions and consequently the EC increases. To reduce the salt accumulation in substrates, the controlled leaching with water of good quality is an alternative. The use of mulching with polyethylene or polypropylene sheet reduces the water consumption, increases the calculated water use efficiency and decreases the EC of the substrate; so the mulching is an alternative to control of EC too.

On the one hand, positive evidences of nutrient solutions reuse are reported, which necessarily involves regulation of the EC. Therefore, recycling and reuse of solutions is a trend in searching for sustainable agricultural production systems. Brun et al. reported recycling systems based on EC control, consisting of adding a water complement to the drainage to decrease the EC and a complement nutrient solution to obtain the desired EC. Carmassi et al.developed a simple model for the changes in ion concentration and EC of recirculating nutrient solution in closed-loop soilless culture on the basis of balance equation for nutrient uptake by hydroponically-grown plants. In this model, the crop evapotranspiration is compensated by refilling the mixing tank with complete nutrient solution.

Recently, an in situ optimal control method of nutrient solution composition has been proposed. Instead of modeling the correlations between greenhouse vegetable growth and nutrient solution, this method is based on Q-learning searches for optimal control policy through systematic interaction with the environment.

Even though, Bugbee indicates that the monitoring ions in solution is not always necessary. In fact, the rapid depletion of some nutrients often causes people to add toxic amounts of nutrients to the solution. Besides, it has been demonstrated the existence of a wide cultivable microbial community in the nutrient solution before recycling and recirculation, which supports the necessity of disinfecting nutrient solutions used in soilless cultivation systems, during the recycling process, in order to ensure crop sanitation and avoiding plant disease spreading.

Temperature Control

The temperature of the nutrient solution has a direct relation to the amount of oxygen consumed by plants, and an inverse relation to the oxygen dissolved in it, as it was previously indicated. Temperature also affects solubility of fertilizer and uptake capacity of roots, being evident the importance of controlling this variable especially in extreme weathers. Each plant species has a minimum, optimum, and maximum temperature for growth, which requires the implementation of heating or cooling systems for balancing the nutrient solution temperature.

The underground water pipe system for energy-saving control of nutrient solution temperature consists of a large-sized pipe filled with water under the ground, and a unit for circulating the nutrient solution between the cultivation bed and the underground water pipe. The temperature condition in the underground water pipe 1.5 m below the ground surface is stable as compared to that in greenhouses which excessively high temperatures in summer and low in winter. During the circulation, heat can be exchanged between the nutrient solution and the water stored in the underground water pipe. Furthermore, this circulation warmed the nutrient solution excessively chilled in cold winter nights.

Nam et al. evaluated the cooling capacities of three different systems, which used either polyethylene or stainless tube in the solution tank, or a counter flow type with double pipes, having 41, 70 or 81% of cooling load in a hydroponic greenhouse, respectively.

pipes, having 41, 70 or 81% of cooling load in a hydroponic greenhouse, respectively. Villela et al. evaluated the cooling of the nutrient solution at about 12°C by using a heating exchange device on the productivity of two varieties of strawberry. The cooling of the nutrient solution conferred better productivity of Sweet Charlie variety; whilst it didn't cause any effects over the Campinas variety.

Deep sea water (DSW) can be used in the preparation of nutrient solutions due to its nutrient content. Likewise, it is one alternative for nutrient cooling systems in hot season due to its low temperature. Cold DSW pumped inside pipe through cultivation bed might decrease temperature of nutrient solution by heat exchange between nutrient solution and DSW. For environment reason, after being used for cooling system, DSW that contained abundant nutrient can be used as nutrient supplement for tomato plant by diluting into standard nutrient solution. It is suggested that DSW might increase fruit quality because of its enrichment of nutrient solution.

Oxygenation of Nutrient Solution

The consumption of O_2 increases when the temperature of nutrient solution increases too. Consequently, it produces an increase in the relative concentration of CO_2 in the root environment if the root aeration is not adequate.

The concentration of oxygen in the nutrient solution also depends on crop demand, being higher when the photosynthetic activity increases. A decrease bellow 3 or 4 mg L^{-1} of dissolved oxygen, inhibits root growth and produces changes to a brown color, which can be considered as the first symptom of the oxygen lack.

Nonetheless, substrates under long cultivation periods usually present increase of organic matter content and microorganism activity, which could lead to an increase of the competition for oxygen in the root environment. Yet, roots are densely matted within the substrate, which alters diffusion and supply of oxygen.

The supply of pure, pressurized oxygen gas to the nutrient solution is an oxygen-enriched method often used for research purposes, and it is called oxyfertigation.

Bonachela et al. evaluated the response to oxygen enrichment of nutrient solution in of autumn-winter sweet pepper and spring melon crops grown on rockwool slabs and perlite grow-bags, compared to non-enriched crops. The pressurized oxygen gas was dissolved in the nutrient solution during each irrigation with a gas injector within the irrigation pipe. The use of inexpensive systems of substrate oxygen enrichment should be restricted to rockwool substrates and to crop periods when a high oxygen demand coincides with low oxygen availability, such as the period from melon flowering phase.

The supply of potassium peroxide as an oxygen generator on vegetable crops grown in commercial substrates once a week was evaluated in sweet pepper, melon and cucumber. Results indicated that the application of potassium peroxide at a concentration of 1g L^{-1} is the best fraction to use in soilless culture. The treatment with potassium peroxide increases the yield of sweet pepper and melon in 20 and 15% respectively, in comparison to the control, whereas there was no significant difference in cucumber yield.

Preparation of Nutrient Solution

Nutrient Solution Design

Hansen indicates that the addition of plant nutrients to hydroponic systems may be performed according to the plant nutrient requirement. Application of nutrients may be performed according to analyses of a specific crop stage that may describe the consumption of the various typical nutrients of the particular crop or by means of analyses of the total plant needs quantitatively adjusted to the rate of growth and the amounts of water supplied. Thus, the composition and concentration of the nutrient solution are dependent on culture system, crop development stage, and environmental conditions.

Likewise, Steiner proposed that in soilless cultures any ionic ratio and any total concentration of ions can be given, as precipitation limits for certain combinations of ions are considered. Thus, the selection the concentration of a nutrient solution should be such that water and total ions are absorbed by the plant in the same proportion in which those are present in the solution.

Steiner developed a method to calculate a formula for the composition of a nutrient solution, which satisfies certain requirements. Later on, he evaluated five different ratios of NO_3^- : anions ($NO_3^- + H_2PO_4^- + SO_4^{2-}$) and three of K^+: cations (K^{++} Ca^{2++} Mg^{2+}), combining also the two groups, resulting in a full factorial design; all solutions had the same osmotic pressure and pH value. In this system, the relative concentration of K^+ increases at the expense of Ca^{2+} and Mg^{2+} concentrations. Furthermore, the ratio 3:1 between Ca^{2+} and Mg^{2+} is constant. Similarly, the ratio H_2PO_4: SO_4^{2-} (1:9) is constant, while the changes in the NO_3^- concentration are produced at expense of the $H_2PO_4^-$ and SO_4^{2-} concentrations.

On the other hand, Van Labeke et al. studied Eustoma grandiflorum responses to different nutrient solutions differing in ion ratios using an experimental as a {3.1} simplex centroid design, one in the cation factor-space and the other in the anion factor-space, which is depicted in figure. Then, De Rijck & Schrevens investigated the effects of the mineral composition of the nutrient solution and the moisture content of the substrate on the mineral content of hydroponically grown tomato fruits, using "design and analysis of mixture systems", a {3.1} simplex lattice design extended with the overall centroid set-up in the cation factor-space (K^+, Ca^{2+} and Mg^{2+}) of the nutrient solution. For each nutritional composition two moisture contents (40 and 80% of volume) of the substrates were investigated.

After this short sample illustrates some aspects to be considered in the preparation of nutrient solutions.

Water Quality

According to Tognoni et al., water quality in hydroponics can be limited to the concentrations of specific ions and phytotoxic substances relevant for plant nutrition as well as the presence of organisms and substances that can clog the irrigation systems.

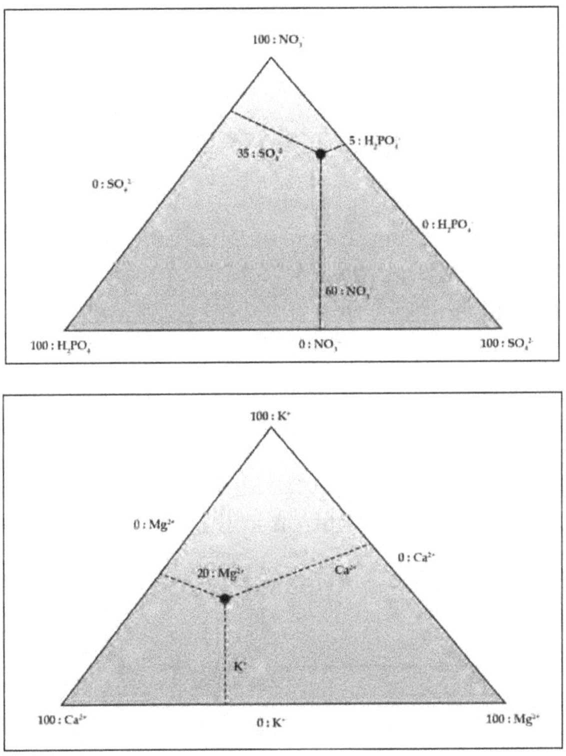

Proposed ratios NO_3^- : anions ($NO_3^- + H_2PO_4^- + SO_4^{2-}$) (a) and K^+: cations (K^{+}+ Ca^{2+}+ Mg^{2+}) (b) by Steiner.

Regarding the presence of organisms both in water for preparing nutrient solution and in recirculating nutrient solution, its control can be achieved by heat treatment, UV radiation and membrane filtration. However, cheaper chemical treatments as sodium hypochlorite, chlorine dioxide and copper silver ionization may partly solve the pathogen problem; with the disadvantages that introduce a potential accumulation of other elements in closed systems.

It is necessary to carry out a chemical analysis of water to be used in the nutrient solution. Knowing the kind and concentration of ions allows identifying on the one hand, those that are needed in the nutrient solution and therefore can be subtracted from the original formulation; and on the other hand, to take decisions about ions not needed in the nutrient solution.

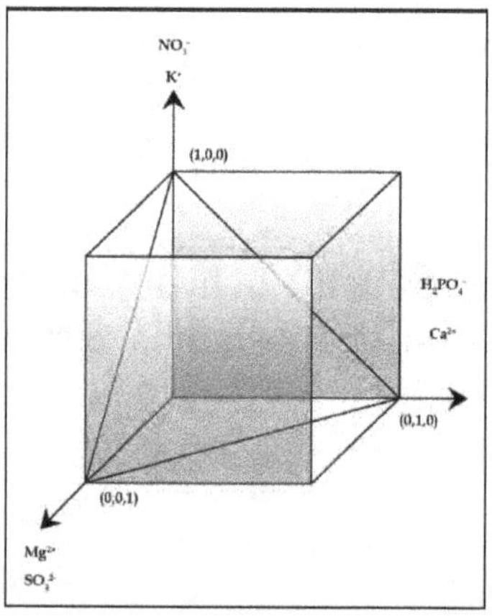

Experimental {3.1} simplex centroid design used by Van Lecke et al.

DSW can be used for preparing nutrient solutions. Chadirin et al. reported that plants treated with circulated water having an EC of 20 dS m^{-1} produced tomatoes with highest soluble solids, 8.0% Brix or increased yield in 30% in comparison to the control. Nevertheless, Pardossi et al. showed that nutrient solution with high EC (9 dS m^{-1}) had no important effects on both crop yield and fruit quality.

Fertilizer Source for Nutrient Solution

Table has a list of commonly used fertilizers and acids in hydroponics, as well as some characteristics of interest for plant nutrition applications.

Table: Fertilizers containing macronutrients that are commonly used in the preparation of nutrient solutions.

Fertilizers	Formula	Nutrient percentage	Solubility, g L^{-1} at 20 °C
Calcium nitrate	$Ca(NO_3)_2 5H_2O$	N:15.5;Ca: 19	1290
Potassium nitrate	KNO_3	N: 13; K:38	316
Magnesium nitrate	$Mg(NO_3)_2 6H_2O$	N: 11; Mg:9	760
Ammonium nitrate	NH_4NO_3	N:35	1920
Monopotassium phosphate	KH_2PO_4	P: 23; K: 28	226
Monoammonium phosphate	$NH_4H_2PO_4$	N; 12; P: 60	365
Potassium sulphate	K_2SO_4	K: 45; S: 18	111
Magnesium sulphate	$MgSO_4 7H_2O$	Mg: 10; S: 13	335
Ammonium sulphate	$(NH_4)_2SO_4$	N: 21; S: 24	754
Potassium chloride	KCl	K: 60; Cl: 48	330

Hoagland Solution

The Hoagland solution is a hydroponic nutrient solution that was developed by Hoagland and Snyder in 1933, refined by Hoagland and Arnon in 1938 and revised by Arnon in 1950. It is one of the most popular solution compositions for growing plants (in the scientific world at least) with more than 15,000 citations listed by Google Scholar. The Hoagland solution provides every nutrient necessary for plant growth and is appropriate for supporting growth of a large variety of plant species.

The solution described by Hoagland in 1933 has been modified several times, mainly to add ferric EDTAs and alter the number and concentrations of micronutrients. In the revision of 1950, only one concentration (Mo 0.01 ppm) was changed compared to 1938, while the concentration and composition of macronutrients remained the same since 1933.

- N 210 ppm.
- K 235 ppm.
- Ca 200 ppm.
- P 31 ppm.
- S 64 ppm.
- Cl 0.65 ppm.
- Na 1.2 ppm.
- Mg 48.6 ppm.
- B 0.5 ppm.
- Fe 2.9 ppm.
- Mn 0.5 ppm.
- Zn 0.05 ppm.
- Cu 0.02 ppm.
- Mo 0.05 ppm.

The Hoagland solution has high concentrations of N and K so it is very well suited for the development of large plants like tomato and bell pepper. Due to relatively high concentrations in the stock solutions the solution is very good for the growth of plants with lower nutrient demands as well, such as lettuce and aquatic plants, with the further dilution of the preparation to 1/4 or 1/5 of the modified solution.

Salts and acids to make up the Hoagland solution formulations:

- Potassium nitrate, KNO_3.

- Calcium nitrate tetrahydrate, $Ca(NO_3)_2 \cdot 4H_2O$.
- Magnesium sulfate heptahydrate, $MgSO_4 \cdot 7H_2O$.
- Potassium dihydrogen phosphate, KH_2PO_4.
- Ammonium dihydrogen phosphate, $(NH_4)H_2PO_4$.
- Iron(III)-EDTA or Iron chelate, Fe-EDTA or Fe-EDDHA.
- Boric acid, H_3BO_3.
- Copper sulfate pentahydrate, $CuSO_4 \cdot 5H_2O$.
- Zinc sulfate heptahydrate, $ZnSO_4 \cdot 7H_2O$.
- Manganese chloride tetrahydrate, $MnCl_2 \cdot 4H_2O$.
- Molybdic acid monohydrate, $H_2MoO_4 \cdot H_2O$.
- Sodium molybdate dihydrate, $Na_2MoO_4 \cdot 2H_2O$.

Component	Stock Solution	mL Stock Solution/1 L
Macronutrients		
2M KNO_3	202 g/L	2.5
2M $Ca(NO_3)_2 \cdot 4H_2O$	236 g/0.5 L	2.5
Iron (Sprint 138 iron chelate)	15 g/L	1.5
2M $MgSO_4 \cdot 7H_2O$	493 g/L	1
Micronutrients		
H_3BO_3	2.86 g/L	1
$MnCl_2 \cdot 4H_2O$	1.81 g/L	1
$ZnSO_4 \cdot 7H_2O$	0.22 g/L	1
$CuSO_4 \cdot 5H_2O$	0.08 g/L	1
$H_2MoO_4 \cdot H_2O$ or	0.09 g/L	1
$Na_2MoO_4 \cdot 2H_2O$	0.12 g/L	1
Phosphate		
1M KH_2PO_4	136 g/L	1

The Hoagland solution formulation based on $(NH_4)H_2PO_4$ instead of KH_2PO_4 must be prepared according to a different protocol, which is referred to in the circulars of 1938 and 1950 as solution. Sprint 138 iron chelate is produced as sodium Fe-EDDHA, while Hoagland's original solution formulation optionally contains ferric or ferrous tartrate but no sodium ions. Other micronutrients (e.g., Co, Ni) and rather non-essential elements (e.g., Pb, Hg) mentioned in Hoagland's 1933 original publication (known as A-Z solutions a and b) are no longer included in his later circulars. These elements and organic compounds are not necessary for normal plant nutrition. As an exception, there is evidence that, for example, some algae require cobalt for the synthesis of vitamin B12. On the other hand, it is evident that the modified Hoagland solutions of 1938 and beyond are balanced nutrient solutions that answer the question how to compose and concentrate the solutions best suited to the growth of plants.

Hydroponic Additives for Plants

The Most Vital Hydroponic Additive: Nutrients

Base nutrients are the foundation of your plant's diet. They are the basic plant food which will provide all the necessary elements for plant growth such as Nitrogen, Phosphorus, Potassium, Calcium, Magnesium, Iron, Manganese, Zinc and some other minor elements. Nutrients are customised for growing methods, for example nutrients for Coco coir have more calcium and magnesium as these elements are absorbed by the coco fibres.

Seaweed/Kelp

The most important hydroponic additive after the base nutrient are Seaweed or Kelp extracts which improve your plants immunity on a few fronts.

First, since it facilitates superior nutrient uptake and provides powerful plant hormones, it will make your plant healthier and more resistant to disease. Second, kelp contains some natural antibiotic properties, which can help in suppressing harmful bacteria.

The extra carbohydrates in seaweed extracts can be invaluable for increasing yield and overall quality.

Root Stimulators

Strong vigorous roots are essential for supporting large healthy plants. Without developed roots the plant will start slowing its growth.

Enzymes

Enzymes break down old matter in your root zone and turn it into sugars and starches. Not only do they create healthy root zones but, they also help the transfer of nutrients to the plant.

Having a healthy root zone is similar to having a tidy work space, easy access to whatever you need, whenever.

Vitamins

Vitamins help a plant feel better when stressed, and keep a plant healthy. Vitamins will help with resistance to fungal rots and insect attacks for instance. While plants manufacture vitamins for themselves, if they have a supply of them, they can turn their energy to producing other elements they need, and thereby speed up growth.

Amino/Acids

The addition of amino acids to a hydroponic nutrient solution often has a dramatic effect on plant growth. Again, the secrets of amino acids lie in nature. As beneficial microorganisms grow and multiply in a healthy, organic soil, they produce enzymes that break down organic matter.

Both humic and fulvic acid are recommended for use in all stages of plant development. They are a great way to supercharge your plants, improve your soil and get more complete use of your nutrients. Give humic and fulvic acid a try if you want to really see what your plants are capable of! There are amino acids in many hydroponic additives.

Cleansing the Water Supply

Have you ever been fighting a cold, not really become sick, but just been a little slowed down? This is what happens to a plant which is fed with an uncleansed water supply. With water free from pathogens (such as pythium) the plant can concentrate on growing rather than fighting off bad microorganisms in the root system.

If you keep your system and root zone really clean you will see an improvement in the plants vigour.

Silica

Consider silica like a missing link between plant vitality, strength, resistance to infection, and increased harvests. The silica helps strengthen cells, and plants are much healthier from the continual addition of silica.

Ensure you have a highly soluble form and you will begin to see the effect from around 2 weeks onwards.

Foliar Sprays

Foliar spraying, although not a substitute for healthy grow media; can be beneficial when a plant is suffering from certain nutrient deficiencies. Foliar plant spray involves applying nutrients directly to a plant's leaves as opposed to putting it in the soil.

Foliar feeding is similar to humans putting an aspirin under their tongue; the aspirin is more readily absorbed into the body than it would be if it were swallowed. A plant takes nutrients through the leaf much quicker than it does through the root and stem. Foliar feeding is generally done in evening or just before lights off, since heat causes the pores on some species' leaves to close.

Flower Boosters/Weight Enhancers

Why is this not the #2 hydroponic additive in the top 10. Flowering boosters are meant to take an already very healthy plant and make it better. If you add flowering boosters to a weak or distressed plant, it is not going to do anything but make things worse.

With proper use and timing boosters ARE invaluable in maximizing flower/fruit size and taste.

An improper balance or excess of nutrients can cause a number of problems including "root burn" or "nitrogen / nutrient burn". This causes the roots and leaf edges to take on a burnt appearance (yellow to black). If you are a grower who is already seeing this problem and are using "flowering boosters" or extra nutrients, you should stop immediately and flush the plants with water to rid the excess.

Mixing Solutions

Hydroponic nutrient products are generally available in the form of a concentrated liquid or a powder at your local indoor gardening store. These days the most popular choice among hobby hydroponic growers are concentrated liquid products probably because most folks find them easier to measure out and work with over the powder products. However powder products are a great option too because they are less bulky and therefore cheaper and more efficient to ship.

Don't confuse hydroponic nutrients with regular garden fertilizer products. They are not the same. Only nutrients specifically designed to work in a soilless environment will work in hydroponics. Regular garden fertilizer products are designed for use with soil, not hydroponics. Don't be tempted to try organic nutrient products in this situation either. Unless the label says it's suitable for hydroponics, it probably isn't.

Okay, so how do you go about creating a hydroponic nutrient solution that's ready to feed to your plants? Well, obviously the first thing you're gonna need to do is mix the nutrient concentrate with water. Manufacturers of liquid hydroponic nutrients generally offer products in the form of 'one-part', 'two-parts' and 'three-parts.' By splitting the concentrated mineral solutions into different components manufacturers are able to offer products in higher concentrations. Remember, one-part, two-part and three-part hydroponic nutrient products basically contain all the same stuff it's just a matter of presentation and usage. Some growers (beginners and experts alike) prefer the ease-of-use of a one part, for example. Other growers enjoy the ability to experiment and adapt afforded to them by the multi-part products.

Unsurprisingly it comes as three bottles labeled 'GROW', 'MICRO' and 'BLOOM'. Generally speaking, you need to use all three parts whatever stage your plants are at just at different ratios and concentrations.

Fill a clean container, keeping track of the amount of water you put in. Don't fill it up too full or you'll make a mess when stirring it later. Here we've got two gallons.

Start with the purest water you can. Good quality water is critical to success when growing using any method, but it's especially important for hydroponics. Fill a clean container, keeping track of the amount of water you put in. Don't fill it up too full or you'll make a mess when stirring it later. Here we've got two gallons. Now, quality water is the most often overlooked component of any garden but it's especially important for hydroponics.

Remember, a hydroponic nutrient solution is a precise blend of different minerals at various ratios. If the levels of any given element become too high it can cause 'lock-out' meaning that other essential elements become unavailable to your plants. So, if you plan to use tap or well water, you should ideally get it tested first to see what's in it. At the very least, take a conductivity meter and test your source water to get an overview of the total amount of dissolved solids in it.

The first part we're going to add is the Micro. Shake the bottle vigorously then check the seal is intact if opening for the first time. The Micro component contains elements such as Calcium, Boron, Copper, Iron, Manganese and Zinc as well as some Nitrogen. The water we're using here is quite hard (around 300 PPMs) so we're going to use a hard water formulation of the Micro part which contains reduced levels of Calcium to counteract the increased residual levels in our water.

Consult the manufacturer's label for dosage instructions. Here, we're going to make a mild general purpose hydroponic nutrient solution for vegetative growth. The instructions say 1 teaspoon per gallon of each part. As we've got two gallons of water, that's two teaspoons of Micro, two teaspoons of Grow and two teaspoons of Bloom.

Using a clean measuring cup, measure two teaspoons (or 10 milliliters) of Micro. Pour into the water. Stir well. Next, take the Grow component. This contains ammoniacal nitrogen, nitrate nitrogen, phosphate, potassium and magnesium. Shake and measure two teaspoons into the measuring cup. Be sure to wash the measuring cup in between each use. It's absolutely vital that the different parts never come into contact with each other in their concentrated form otherwise they won't form a solution properly that's why they're kept separate! Pour this into the water and stir well again.

Finally, take the Bloom component. This contains phosphate, potassium, soluble magnesium and sulfur. Shake, measure out a teaspoon per gallon, pour into the water, and stir once more. There you now have a hydroponic nutrient solution containing all the elements your plants need for growth and bloom.

Oh but hang on a second there's one important final step! In order to optimize the availability of all these elements, you need to check and, if necessary, adjust the pH of your nutrient solution. Don't worry it's real easy to do. First, you'll need to measure the pH. Here we're using a calibrated digital pH meter. Insert it into the nutrient solution to take a reading.

Our target is anywhere between 5.5 and 6.5. This is the best range to ensure the broadest spectrum of availability. As you can see, the pH is too currently too high so we're going to have to add some pH adjuster to lower the pH.

Don't add pH adjusters directly to your nutrient solution. The sudden drop or rise in pH can actually damage your nutrient solution by causing elements to precipitate out of it. Instead, make a dilute solution by adding a pipette's worth of pH adjuster to some water like this. Always wear gloves as pH adjusters are corrosive and can damage your skin. Making up a dilute solution not only makes this process a lot safer but it's also easier to fine-tune the pH of your nutrient solution.

Keep adding, bit by bit, stirring, and re-testing the pH until it falls within the desired range. Finally, ensure your nutrient solution is at the right temperature. Around 64 to 66 degrees Fahrenheit is ideal. It should feel tepid (neither warm or cold) to the touch. Plants don't like taking icy baths any more than you do. If your nutrient solution is too warm, dissolved oxygen levels will be too low. Ideally, use a nutrient solution thermometer to regularly check its temperature.

If you are managing a hydroponic garden, you have two options when it comes to your nutrients: You can purchase a premixed nutrient solution or you can mix your own at home. Buying a premixed nutrient solution is a quick fix that works well. However, mixing your own fertilizer allows you to fine-tune the level of each nutrient to your plants' exact requirements, and you can save a ton of money in the long run. Also, making your own hydroponic nutrient solution is easy. It just requires some preparation and a keen eye.

Before you mix a drop, you need to take the time to read about nutrients. Find out what they do and how they work. Also, learn about the types of nutrients that your individual plants need. Each nutrient corresponds to a different function within the plant. All plants depend on the same primary macronutrients and micronutrients, but their ratios in solution will vary based on what you are growing. Know your plants, and know what and how much they need before you begin.

Next, you're going to need supplies. The most important things on your shopping list are the various nutrient salts that will serve as the base of your fertilizer. Phosphorus, nitrogen, calcium and all the other essential elements you need are available in this solid crystalline form.

When the nutrient salts are mixed with water to make your solution, they break down and provide nutrition as they are absorbed by your plant's root system.

Your water should be clean. You always end up with a better feed mix if you can use water run through a filtration system to remove unwanted contaminants. Next, you need a few buckets; one for each part of the solution.

For example, if you are using an "A, B, Bloom" system, you need three buckets. Also, buy a digital scale that can measure down to the hundredth of a gram, as your measurements will need to be precise.

Next, you need your typical measuring spoons and cups. Don't forget to use rubber gloves. Safety is paramount, so always wear gloves when handling nutrient salts or any other chemicals.

Now that you have your supplies, fill your buckets with the appropriate amount of water. Next, use your measuring cups, spoons, and scale to weigh out the proper amount of salts. Prepare each nutrient on its own, and do it carefully.

Pouring your Salts

Salts dissolve almost immediately after being introduced to water. Pour them in slowly and disperse them around the bucket, trying to avoid unnecessary splashing. After one salt has dissolved, you can proceed with the next.

Once you have measured out and introduced all of your salts, put a lid on the bucket and give it a good shake to break up any clumps that may have been missed.

Now that you have your preliminary mix done, you need to check and adjust your pH. This is of utmost importance. Plants require a pH in a range of about 5.5 to seven, with most falling under 6.5. With water being a neutral medium (that is, having a pH of seven), you are going to need some pH down solution.

Adjusting pH

Solutions designed to reduce pH are highly acidic, meaning a little bit goes a long way. Make a separate diluted pH down mixture by using just a few drops of pH down concentrate with a gallon of water.

This gives you a solution with a pH of around two, give or take. Then, slowly add the diluted pH down mixture to your nutrient mix, stopping to check your result frequently.

Once you get your pH in line, it's time to fine-tune your electrical conductivity (EC). That is, you need to check if the ratio of total nutrients to the amount of water is correct. An EC measurement is similar to a part per million (ppm) reading, but it has become the new, more accurate standard in measuring nutrient ratios.

There are several ways to get your EC reading, but the easiest is to use an electronic EC meter. You want your EC anywhere from 0.8 to three, those being the extremes. Most gardening applications require an EC of somewhere between 1.5 and 2.5.

Above that, you risk a nutrient burn. If your EC is too high, add enough water to bring your EC down to the desired range.

Nutrients for Hydroponic Systems in Fruit Crops

Hydroponics is a technology in which nutrient uptake occurs through plant roots dipped in the nutrient culture. Prior to use of hydroponic culturing, the crop physiological functioning system must be clear. For optimum crop growth and functioning, mixture of sunlight, carbon dioxide, water, and nutrient elements for photosynthetic efficiency is needed. Besides, the minerals are either found naturally in the soil or supplied through the fertilizers to the soil. Thus, it is evident that plant needs mineral derived from the soil for its growth, not the soil which becomes a basic idea behind the development of hydroponics. Moreover, the roots also need an optimum supply of oxygen for uptake and transport of metabolites to the whole plant. In hydroponic system, the roots of plants are in direct contact with the nutrient solution only vis-a-vis the absorption of nutrients occurs more easily than soil grown plants. This system also performed the crop plants with faster growth and higher yields and in turn saves energy for extensive root system development and thus, more of energy can be diverted toward leaf and stem growth. Besides, it also offers significant advantages over traditional farming, producing greater yields, faster growth, and possibly year-round crop production. This system further allows recycling of nutrient solution without wastage of water. The amount and composition of the nutrients to which plants have access can be monitored precisely by the grower. It also allows the grower to control the pH level of the solution and protects plants from pests and diseases. Seedlings of fruit crops can also be grown hydroponically in earlier stages of growth and then transplanted to the field. Different fruit crops namely, grapes, raspberry, blackberry, blueberry, and strawberry can be grown hydroponically. Strawberries and blueberries perform best under hydroponics system because of acidic soil requirement are best suited to blueberries and be grown under hydroponic system as pH level and nutrient content are easily controlled and maintained in hydroponics.

Hydroponics Technique

There are different types of hydroponics technique being employed for growing of plants. There are mainly three types of hydroponic systems.

Nutrient Film Technique

It is a system in which the nutrient solution is passed through the roots of plants placed in a channel. The plants are placed on channels made up of wood, rigid to flexible tubes or plastic and the nutrient rich solution is either pumped through it or passed under gravitation reaching the root system effectively. This technique is effective only for the plants with large root system. The nutrient film technique (NFT) was developed during the late 1960s by Dr. Allan Cooper at the Glasshouse Crops Research Institute in the

U.K. NFT is the growing of plants, bare-rooted in long, waterproof channels, down which flows a very shallow stream of re-circulating water, into which are dissolved all the minerals required to grow healthy plants. NFT is a hydroponic technique wherein a very shallow stream of water containing all the dissolved nutrients required for plant growth is re-circulated past the roots of plants in a watertight gully, also known as channels. According to the pre-requisite to achieve a nutrient film situation more effectively is described as (i) ensuring the gradient down where water flows is uniform and not subject to localized depressions, (ii) rapid inlet flow rate that a considerable depth of water flows down to the gradient, (iii) adequate width of the channels to avoid any damming up of the nutrients, and (iv) flat channel base but not curved due to which otherwise will be a considerable depth of liquid along the center of a channel with a curved base. N.F.T. system is fairly a simple design. However, this is not best suited for smaller quick growing plants.

Deep Film Technique

It is a technique in which the plants are grown with roots submerged in floating nutrient solutions (10–20 cm deep) on a flat table. This method is effective for growing plants with short root system and is relatively cheaper.

Substrate Method

In this method, the plant growth is supported by using materials such as stones, vermiculite, perlite, etc. in structures such as tubes and pots. This system results in affective utilization of nutrients and reusing of the nutrient solution as the drained solution is re-circulated in the system. This system is commonly utilized in Asia, Europe, and Israel for strawberry cultivation by using several trough systems.

Among the three systems, NFT and DFT are the most commonly used methods. The basic idea behind the working of NFT is the recycling of the nutrient solution. This technique offers major advantages over other systems like low cost of installment, easy operation and management as well as conservation of nutrients and water. In this system, nutrient solution enriched with material like sand, vermiculite or rock wool is passed and re-circulated through a slope consisting of plants placed in a plastic trough. It provides the optimum amount of nutrients to the plant through its root system. It is best suited for short period crops like lettuce, strawberry, and raspberries. Nutrient film technique has been termed as a promising tool in the areas with limited land resources.

Various materials that can be used for hydroponics in fruit crops include that of mineral origin and organic origin. Vermiculite is one of the mineral substances utilized in pears, peach, and tangerine seedlings as it is free of pathogens attack and have high water retention capacity. In case of grapevine, sand because of its easy acquisition is used as it results in the increased absorption of macronutrients. Rice hulls have been found to be effective organic source in growing strawberry plants under hydroponics.

For strawberry, perlite or vermiculite is the best growing medium. Materials like coconut coir or peat moss should be avoided as they absorb too much of nutrient solution and cause condition of suffocation to the plants.

For growing strawberries through hydroponics, two systems viz., "closed" and "open" are employed. Among these two, recirculation of nutrient solution occurs in the closed system with plants grown in channels or pipes. The closed system can have continuous nutrient supply or the supply can be at irregular intervals when the plants are grown in pots. In case of pot grown plants, different substrates having high water retention power help in supplying nutrients to the plant system. Whereas in case of open system, there is no recirculation of nutrient solution and is applied to the plants with the help of drippers.

In strawberries, NFT is the most commercially practiced method. In this method, the runners of strawberry are placed in net pots in which the roots are covered with clay medium in the root zone, which help in increasing the strength of plant. They can also be placed with plugs into the net pots which are framed in the NFT channels. It must be placed such that there is continuous contact with the flow of nutrient solution in the initial weeks of root development. The optimum amount of oxygen must also be maintained during the operation of system with 14–16 hours of daylight. The nutrient discharge should be 1–2 liters/minute with circulation pump running all the time. In NFT, it must be grown in low humidity conditions and care should be taken as it is susceptible to root rot.

Nutrient Solution

Nutrients are the basic elements for hydroponics, and nutrient solution is the liquid fertilizer solution prepared in definite composition to support plant growth. The plants need is fulfilled through the ionic form of nutrients with proper oxygen supply and temperature. Environmental factors and nutrient solution are the two important factors to be considered for productivity in hydroponics. Supply of nutrient elements depends upon the requirement of crop, and the frequency of application is based upon the type and age of crop, the type of material used in media and the prevailing environmental conditions.

The kind of nutrient solution varies according to crop species, their growth stage, environment, and other related factors as there is no ideal nutrient solution available to meet the needs of all the crops. Among fruit crops, a lot of researches have been done regarding the nutrient solutions. In grapevine, macronutrient absorption based on nutrient culture of was studied. It was reported that higher accumulation of nutrients resulted in increased vigor of rootstocks Jales, Tropical and Campinas. Solution was also used in pineapple cv. Perola produced through micropropagation in hydroponics system. Long Ashton nutrient solution was used in grapevine under hydroponics. The nutrient solutions for some fruit crops such as peach and pear have not been disclosed.

Composition

The composition of various nutrients in the nutrient solution plays a major role, as the uptake of these nutrients in optimum amounts affects the functioning of plants, thereby affecting its growth. Testing of water must be done before using it is in hydroponics for nutrient solution to get the accurate details about the properties of water. In strawberry for a closed type of NFT, the nutrient solution with the following composition can be used:

Nutrient elements	Quantity (ppm)
Nitrogen (nitrate form)	160.0
Nitrogen (ammonium form)	15.0
Phosphorus (PO_4)	50.0
Potassium	210.0
Calcium	190.0
Magnesium	50.0
Iron	6.0
Boron	0.50
Manganese	0.50
Copper	0.10
Zinc	0.08
Molybdenum	0.05

There are different nutrient solutions being standardized containing different concentrations of nutrient elements. Hoagland and Arnon nutrient solution has been used for the production of seedlings of guava and pineapple. Furlani et al. nutrient solution was used in production of guava seedlings. Yamazaki solution can be used for strawberry which includes N(NO_3:5; NH_4:0.5); P:1.5; K:3; Ca:2; Mg:1; S:1; Fe:3; B:0.5; Mn:0.5; Zn:0.05; Cu:0.02; Mo:0.01 in meq/L.

Nutrient elements	Hoagland and Arnon	Furlani et al.	Hewitt	Cooper
	mg/L			
N	210	202	168	200–236
P	31	31	41	60
K	234	193	156	300
Ca	160	142	160	170–185
Mg	34	39	36	50
S	64	52	48	68
Fe	2.5	0.26	2.8	12
Cu	0.02	0.04	0.064	0.1
Zn	0.05	1.8	0.065	0.1
Mn	0.5	0.37	0.54	2.0
B	0.5	0.06	0.54	0.3
Mo	0.01	0.11	0.04	0.2

Nutrient Preparation

The nutrient solution for hydroponics can be either bought premixed or can be prepared by self. Plants require same macro and micronutrients but in different ratios. So, the nutrients supplying fertilizers must be bought based on the plants' need as each nutrient has distinct function in different plants. For the preparation of nutrient solution, the nutrient fertilizers are mixed with water which breaks downs to release nutrients. The selection of fertilizers should be such that the amount of water and nutrients present in the solution are equal to the amount of water and nutrients taken up by the plants. The nutrient solution preparation requires water of good quality which is contamination free for which chemical analysis is important. The formulations must be based upon the targeted crop and must supply all the essential nutrients. The nutrient solution preparation also requires maintaining optimum levels of pH and EC. For the proper growth of strawberry plants, the pH of nutrient solution used must be between 5.5 and 6 and the ideal EC range is 1.8–2.0 dS/m during growth period and 1.8–2.5 dS/m during fruiting stage. The pH of the solution can be maintained through potassium hydroxide used to increase pH or through phosphoric acid to lower down the pH. At EC higher than 1.2 dS/m to prevent the damage to strawberry plant root area of plants can be flushed with clean water for the removal of accumulated salts.

Recycling of Nutrient Solution

To meet the requirement of plant without any loss of nutrients, the closed system of hydroponics offers a huge benefit for recycling of nutrients reducing economic as well as environmental costs. In a closed system, the water and nutrient supply is equal to their quantity taken up by the plants. This system provides controlled nutrient supply with minimal leaching losses and reduced environmental contamination. There is a continuous supply of nutrient solution touching the roots of the plant which after passing down is recirculated and is again available for the use of the plants. Among the added fertilizers, only 50% is utilized by the plant and 70% of the added water is utilized by the plant for its proper growth and transpiration. Recycling of nutrients through closed system of hydroponics and nutrient film techniques is very efficient as it uses only 10% of the water and 25% of fertilizer to that of conventional systems. The mineral content of the added fertilizers may be reduced due to the uptake by plants which may be replenished from time to time. The recycling and reusing of nutrients and water having huge advantages also poses some issues. There is increase in EC of the nutrient solution if water uptake is greater than the nutrients and reduced EC due to greater nutrient uptakes which as a result disrupt the recycling mechanism of the system. There may be problem of increased concentration of salts, toxic ions, and pathogens in the nutrient solution where recycling of pathogens occurs along with the solution in the system resulting in their build up. The problems being faced in the closed system can be removed by using ultraviolet treatment, heat treatment, and slow sand filtration. It has been proved that among the previously mentioned, slow sand filtration is best as it is chemical free, easy to maintain, and energy efficient with adaptability in components.

Bio sand filter is used in the system against pathogens of Pythium spp. and Phytophthora spp. In case of open system, the nutrient solution is not recirculated but released into the environment after the crop cultivation. It has been proposed that the nutrient solution released to environment can be recycled to be used as irrigation water without by reducing further pollution chances.

Frequency of Application

For growing strawberry hydroponically, the plants must be fed with nutrient solution daily and best time being 6:00 am–8:00 am. The application must be such that conditions like overwatering and drying not occur. For the early stages of plant growth when the plant is small, less amount of nutrients are required. In actively growing period and during summers, large amount of water is consumed by the plants due to increased transpiration rate. The water requirement also varies depending upon the environmental conditions maintained. The amount of nutrients to be added to the system must be based upon the crop usage. The quantity of water and nutrients taken up by the plants can be known by measuring the EC daily. Lower amount of EC indicates more nutrient uptake, and higher level indicates increased water uptake. Based upon the EC levels, the water must be added to avoid buildup of salts. The reservoir must be filled with water once it is lower than the required volume and must be checked for EC and pH. If the EC and pH of the solution is not stable as per the requirement, it must be adjusted accordingly. The refilling of reservoir from time to may result in imbalance of nutrients in the nutrient solution. So, it must be dumped after a period of time to avoid any interruption in the growth of pant. If the reservoir is small, the solution must be dumped after every 10–15 days while in case of large reservoirs once a month.

Nutrient need through Hydroponics

In soil grown plants, the fertilizers are added into the soil or applied through foliar application, but in hydroponics, a solution of ionic compounds helps in delivering nutrients into the plant system. Hydroponics is better in meeting plants nutrient need as under the manipulated set of conditions nutrients are directly supplied to the roots by coming in contact with them. Hydroponics, due to their better control over the environmental conditions, has been proved to be superior and sustainable for growing of different crops. Particularly in case of berry crops like strawberry, it has turned to be a very effective method producing fruits of superior quality with high yielding potential. Hydroponically, grown strawberries have been found to produce fruits with higher amount of vitamin C, vitamin E, and total polyphenols. By following different systems in hydroponics and different substrates, the nutrient demand of the plant can be met more efficiently. The supply of optimum amount of nutrients and water must be ensured depending upon the crop need so that the plant continues to grow without any lack or excess of both nutrients and water. With several discussed advantages, hydroponics is better choice over conventional methods to produce fruits with reduced water and fertilizer use.

pH of Hydroponic Nutrient Solution

The pH levels in small hydroponic systems can often be overlooked if a grower is focusing more on monitoring a solution's electrical conductivity or TDS level, balancing nutrients, providing beneficial additives and avoiding algae and plant pathogen problems.

However, overlooking pH control can be perilous for plants, particularly those that rely on water supplies with high alkalinity.

The pH of the nutrient solution is a major factor in determining the uptake rate of many essential nutrient ions.

Run pH too high and the dreaded nutrient lockout looms. Often, the first sign that pH has drifted out of range is a slight paling or yellowing of the younger foliage as the plants struggle to take up certain essential nutrients.

Many inexperienced growers tend to misinterpret this so a quick check of pH is always worthwhile when troubleshooting growth problems.

In simplest terms, pH is a measure of the acidity or alkalinity of a solution. The pH scale is logarithmic, which means a pH of 4 is 10 times more acidic than a pH of 5 and 100 times more acidic than a pH of 6. With a pH of 7 being neutral, such as with pure water, values below 7 are acidic and those above are alkaline (or basic).

Plants grown in hydroponics have a different optimal pH level than those grown in soil, so soilless gardeners need to be careful not to apply the pH recommendations for soil-grown crops to those they produce in hydroponics.

For most commonly grown hydroponic crops, an optimal pH range is between 5.5 and 6.5. Commercial growers often use a narrower range of 5.8 to 6 for most crops when they are using automatic controllers that regularly dose acid into recirculating systems to maintain this precise level.

The optimal acidic pH range for hydroponic crops is important as it affects the solubility, availability and uptake of several of the essential plant nutrients.

If the pH drifts too high (past 7), plant uptake of some nutrients becomes less efficient. For example, plants can become iron deficient, even if sufficient iron is present in the nutrient solution.

Calcium is also affected by high pH, forming insoluble salts that precipitate out of the nutrient solution to form whitish deposits on reservoir walls, channels and equipment otherwise known as scale.

This type of nutrient lockout will typically show up as iron deficiency on new foliage (yellow, interveinal chlorosis on the foliage) and as tipburn and leaf cupping, which are symptoms of a reduction in calcium availability and uptake.

Inexperienced growers may misinterpret these nutrient deficiency symptoms as a problem with the nutrient formulation or product itself, rather than an issue with pH rising above optimal levels for iron and calcium uptake.

What causes pH Fluctuations in a Garden?

In a healthy, well-run hydroponic system, pH fluctuations are normal and in some instances, such as recirculating nutrient film technique with a large crop of mature plants and small nutrient volume, pH changes can be quite rapid and require frequent adjustments to stay within a narrow range for optimal nutrient uptake.

As plants remove nutrient ions from the solution, the solution's pH drifts up or down. If left uncontrolled, the pH will often drift downwards for several days after planting a new crop, after which the pH will steadily increase.

This is due to the differential uptake of ions from the solution, with the release of hydrogen (H^+) or hydroxyl (OH^-) ions from the root system.

As positive ions such as cations $Ca2^+$, K^+, Mg^{2+} are removed from the solution, hydrogen ions (H^+) are released from the root system to equalize the ratio of anions to cations in the root zone and this lowers the pH of the solution.

When the crop begins an active growth phase, anions such as NO_3 are taken up, which increases the pH through the release of hydroxyl ions (OH^-) into solution.

Once plants are well established, most hydroponic systems tend to see a gradual and continual increase in pH over time, which is countered with doses of diluted acid.

Water Supply and Water Treatment for pH

Many of the pH problems that growers encounter originate from the water supply. Many source waters, including the city supply, can have issues with alkalinity and while that is not a problem for human consumption, it can cause problems in a hydroponic system.

The alkalinity of a water supply describes the strength of a high pH so a water supply with a high starting pH and a high alkalinity takes a far greater volume of acid to bring it down to optimal levels than a water supply with a low alkalinity and of the same pH.

Alkalinity is usually provided on water analysis reports from city supplies and is worth checking if continual pH rises requiring large doses of acid is a problem.

High alkalinity is considered greater than 300 mg/L of calcium carbonate and low alkalinity is less than 100 mg/L. When high-alkaline water is first added to the nutrient reservoir, it can take large volumes of acid (pH down solution) added over a number of days to finally bring down and stabilize the pH in the 5.8 to 6 range.

However, adding large amounts of acid is not only time consuming in terms of monitoring, adjusting and readjusting, but it can also create imbalances in the nutrient ratio as acids also add in nutrient ions.

Nitric, phosphoric and sulphuric acid all add N, P or S to the carefully balanced nutrient solution, so accumulation can occur. Growers with a water supply of high alkalinity can prevent this issue by pre-acidifying the source water down to a pH of 6 before using it to make up nutrient solutions or adding as top-up water to a nutrient reservoir.

Once the water supply has been stabilized at a pH of 6 and the alkalinity countered to the point, it remains at that pH for 24 hours and can be used in the hydroponic system. Much less acid will then be required for pH control.

Hydroponic Nutrient Formulation and pH

Nutrient formulations or products all have different starting pH values because individual salts become more or less acidic when dissolved into water. Salts such as monopotassium phosphate lower the pH more than salts such as calcium nitrate. Most formulations will result in an initial pH of around 5.5 to 6, which is ideal for the growth of hydroponic crops.

In hydroponic solutions, some salts can be used to influence the pH control of the nutrient solution, reducing the requirement for acids during growth development phases of the crop.

Ammonium nitrate is one salt used for this purpose, and the optimal amount is that which provides 10 to 15% of the total nitrogen of the formula in the ammonium form.

Ammonium in nutrient solutions tends to be acidifying because, unlike nitrate, it is a positive ion, and when taken up by plants it is replaced by hydrogen ions, reducing pH in the root zone. In addition, ammonium forms ammonium hydroxide and hydrogen ions that produce a mild acidifying effect when in solution.

Many of the nutrient products on the market designed for hard or alkaline water sources use a certain percentage of ammonium nitrate to help counter pH increase.

While this is effective in keeping pH levels down, the ammonium form of nitrogen can cause growth issues and even toxicity if the percentage used is too high and under certain growing conditions.

Ammonium competes for calcium uptake, so only small levels 10 to 15% of total nitrogen of ammonium as a nitrogen source are recommended for most crops and well-formulated nutrient products usually contain less than this.

Testing and Adjustment pH Levels

The pH of a nutrient solution can be easily adjusted with the use of specifically designed pH up and down products that are diluted and ready to use.

Commercial growers usually make up their own pH adjustment solutions of 10% nitric or phosphoric acid or potassium hydroxide for increasing pH.

By using a ratio of 50% nitric and 50% phosphoric acid, the solution can be kept in better balance as both N and P are added, but when a lot of acid is being used, the nutrient formation should be adjusted for these additional N and P sources, as they are macronutrients.

There are organic acids such as citric acid and white vinegar (acetic acid) that are sometimes used in small systems to bring pH down, but these are weak acids and only give a short-term pH reduction.

Organic acids also add carbon to the nutrient solution, which microbes (both good and bad) love to feed on, encouraging an unwanted reduction in dissolved oxygen in the solution.

There is a wide range of pH testing equipment available, from inexpensive liquid test kits and strips sold by aquarium/swimming pool suppliers to high-tech, electronic meters. Test kits and strips should only be used for the lower pH range many of these measure pH 5 to 7.

Growers with electronic meters should also invest a few dollars in a cheap liquid pH test kit or strips just to double check the accuracy of their electronic meter from time to time. Meter probes tend to drift over time and need to be calibrated on a weekly basis, cleaned and replaced annually.

A quick check against a pH test strip will soon show up any irregularities. Calibration of pH meters is essential and often overlooked by many growers electronic pH meters usually come with buffers.

A pH 4 and a pH 7 solution can be used to check and adjust the meter's readout.

Additionally, buffer solutions for this purpose are usually available through hydroponics retailers and are a good investment.

Always follow the instructions for your pH meter. Some require the probe to be stored wet, others don't. Cleaning the sensitive probe tip is also important for accuracy.

Different Hydroponic Systems

With a recirculating system such as nutrient film technique, checking pH is simple as the solution in the reservoir is what is in direct contact with the roots.

For media-based systems, which are often drip irrigated and in which the nutrient solution may not recirculated, pH control is slightly more complex.

For these types of systems, pH (and EC/TDS) needs to be measured both at the nutrient supply reservoir and in the solution that drains from the base of the grow beds, bags or slabs containing the plants, called the leachate solution.

With large, rapidly growing plants, there will usually be slight differences in pH between the feed nutrient and the leachate solution pH changes as the solution flows past the roots and has ions absorbed/released and the grow medium may influence pH as well.

Adjustment of pH at the nutrient supply reservoir should be based on the pH measured in the leachate draining from the plant's root system.

For example, if the pH is 6.9 in the leachate and 6.2 in the feed solution, the pH of the feed solution should be acidified down to the point where the drainage solution starts to come down to 5.8 to 6 as this is the pH that the root system will be experiencing.

Summing it Up

Measuring and adjusting pH in a nutrient solution is usually quick and simple. There is a wide range of pH testing tools and adjustment products on the market to choose from and retailers can provide valuable advice on what is most suitable for different types of systems.

Keeping a garden's pH within the correct range means your plants will have access to the full menu of essential nutrient ions for optimal growth and productivity.

References

- The-top-10-hydroponic-additives: aquagardening.com.au, Retrieved 10 June, 2020
- How-to-mix-a-3-part-hydroponics-nutrient-solution, plant-nutrition-and-growth-enhancers: just4growers.com, Retrieved 23 August, 2020
- How-to-make-the-perfect-hydroponic-nutrient-solution-3239: maximumyield.com, Retrieved 16 May, 2020
- Nutrients-for-hydroponic-systems-in-fruit-crops: intechopen.com, Retrieved 25 July, 2020
- Perfecting-ph-1212: maximumyield.com, Retrieved 13 March, 2020
- Potassium-nutrition-in-fruits-and-vegetables-and-food-safety-through-hydroponic-system, potassium-improvement-of-quality-in-fruits-and-vegetables-through-hydroponic-nutrient-management: intechopen.com, Retrieved 24 January, 2020

4

Hydroponic Plants

Basil, rosemary, lettuce, spinach, strawberries, bell peppers, herbs, rose, lavender, green beans, tomatoes, etc. are some of the plants that grow best using hydroponic systems. This chapter has been carefully written to provide an easy understanding of these hydroponic plants.

Flowers

Growing flowers lends itself beautifully to hydroponic gardening as they can be grown in larger numbers, and can be grown year-round. Most flowers will do well in a hydroponic garden, and when seedlings are big enough, flowers can be cut or transplanted.

Herbs

Many herbs will grow very well in a hydroponic setting. Some that do the best include anise, basil, catnip, chamomile, chervil, chives, cilantro, coriander, dill, fennel, lavender, marjoram, mint, oregano, parsley, rosemary, sage, tarragon and thyme.

Anise

Anise is a feathery annual that grows from 1 to 2 feet high, has finely cut serrated leaves and very small, whitish flowers in flat clusters. Both the leaves and seeds have a warm, sweet licorice taste. It grows rapidly from seed and should be planted after all danger of frost has passed. The green leaves can be cut whenever plants are large enough and seeds may be gathered 1 month after flowers bloom. Anise leaves can be used in salads and as a garnish; the seeds flavor confections such as cakes and cookies.

Basil

In a protected environment, growing basil can be accomplished throughout the year. Once mature, it can be harvested and trimmed weekly. It responds extremely well to hydroponic growing.

Cannabis

A psychoactive herb derived from the flowering tops of hemp plants. Cannabis is controlled under Schedule I of the Controlled Substances Act of 1970. It is also called bhang, ganja, grass, hashish, marijuana, pot, reefer, tea, and weed. It thrives and grows to a more vigorous plant in a hydroponic system. Most cannabis plants cultivated in the United States begin to flower by late August to early October and the plants are harvested from October to November.

Catnip

It grows well in hydroponics in full sun or partial shade. It is a perennial herb of the mint family that will grow from 3-5 feet tall. It is easily propagated by seed, stem cuttings, or rootball division. Seed should be sown late in fall or early in the spring. Sow in plugs or seeds 12 to 15 weeks before sale. Seeds to finished plugs, 8 to 10 weeks; plugs to saleable plants, 3 to 5 weeks.

Chamomile

It grows well in soilless culture. It is used to make chamomile tea and possesses a number of medicinal qualities. Chamomile grown outdoors prefers full sun but will tolerate some shade. From seeds to finished plugs, 6 weeks; plugs to saleable plants, 4 to 6 weeks.

Chervil

Chervil is a low light cool temperature crop. At temperatures above 70-75, the plant will grow slowly and bolt to flower at an early age. Cool roots are critical to ensure good growth. Unless shade and special cooling is used, chervil is difficult to grow in the summer. Growing time to harvest is 4 weeks. It is the perfect winter crop. Best grown under low lights.

Chives

Chives (a relative of the onion family) require little space and produce a continuous supply of seasoning for salads, and main dishes. Plants are hardy, tolerate a variety of growing conditions, are very aromatic and coveted by gourmet cooks.

Cilantro

Cilantro (a relative of parsley) is a very successful hydroponic herb. It tolerates various pH conditions and lower light conditions. It requires little maintenance and reaches harvest stage within six weeks. It must be regularly trimmed or it will go to seed.

Coriander

Cilantro is a parsley-type herb. When grown for leaves only, it is called cilantro but if allowed to continue growing to seed, it becomes what is called coriander. Coriander requires cool temperatures or it will bolt. Coriander grows best in full sun. Only the immature leaves are sold. (The mature "feathery" leaves that precede bolting are not desirable). To grow in a slab: upon final harvest, pull out plants by the roots and re-seed directly into slab, or transplant. Harvest by cutting tops to crown and allowing to re-grow or pull up by the roots.

Dill

Dill is a flavorful addition for hydroponic herb gardens and it produces new growth when harvested. Replacing spent plants with new seedlings every three to four weeks will insure a constant supply of dill. The compact fernleaf variety produces an abundance of lush growth thereby allowing for numerous cuttings from a single plant.

Fennel

Fennel is a perennial (but usually grown as an annual) that grows to about 3 to 4 feet tall. The leaves are finely divided into thread-like segments and are light green and resemble Dill. It grows easily from seed planted in spring and grows best in full sun. Plants need to be staked when 18 inches tall. Harvest seeds when ripe. Tender flower stalks just before they blossom are best for eating. Fennel seeds are used as a condiment. The leaves have an anise-like flavor and the stems can be eaten like celery. Seeds can be used in cheese spreads and vegetable dishes.

Lavender

Common name for a genus of fragrant herbs or shrubs that is grown as a decorative plant in the garden and for its sweet scent when dried. The dried flowers are used to fill sachets and to perfume clothing or linens. Commercially they, and the green parts, are used for making "oil of spike," aromatic vinegar, and lavender water. True lavender is not fully hardy in northern gardens. It is found more in the milder Pacific Coast and in the South. As seed produces variable plants, propagation is commonly by cuttings of selected plants. Sow seeds 20 to 26 weeks before sale; seeds to finished plugs, 12 to 16 weeks; plugs to saleable plant, 8 to 10 weeks.

Marjoram

A perennial herb usually grown as an annual for its fragrant foliage used to flavor dressings and meat dishes. It thrives in full sun. Since the seed is very small, it is best sown in flats in the greenhouse and the seedlings transplanted outside after all danger of frost has passed. Seeds to finished plugs, 6 weeks; plugs to saleable plant, 2 to 4 weeks.

Mint

Though they are usually grown on land, most mints are actually aquatic or semi-aquatic plants, doing their best in shallow water at the edges of ponds and streams. One of its adaptations to water is to spread by stolons that move out along the top of the soil or just under the surface. Spearmint, peppermint, and orange mint are three examples of mints that grow best in or close to water. Mint grown in water tends to have bigger, more lush foliage than land-grown mint, so it is ideal in hydroponic gardens.

Oregano (Wild Marjoram)

Oregano is a hardy perennial that has sprawling stems that can grow to 2 feet tall. It has small pink or white flowers. Oregano can be propagated by seed or division. Stimulate foliage by cutting back flowers. Replant when plants become woody in 3 to 4 years. Leaves can be used fresh or preserved by drying. Oregano leaves are used extensively as a flavoring on pizza, or sprinkled over lamb. Used generously in Italian-type sauces.

Parsley

Parsley, is an herb well suited for hydroponic growing. Since it has a long tap root, the hydroponic container should be at least 12 inches deep for the best results.

Rosemary

A hardy evergreen sub-shrub grown mainly for its aromatic leaves that are used in culinary seasoning and yield an oil once used in medicine. Small light blue flowers are borne in April or May. The foliage is white and woolly on the underside and dark and shiny above. Plants can grow to a height of 6 feet and last for years but need protection from the cold. It prefers alkaline soil and full sun, but does tolerate moderate shade. Sow in seed flats 22 weeks before sale in 10 cm diameter pots. Seeds to finished plugs, 12 weeks; plugs to saleable plants, 10 weeks.

Sage

Common name for the hardy sub-shrub that is extensively grown for seasoning dressings used with rich meats, and for flavoring sausages and cheese. In hydroponics, it can be grown from seeds protected from cold and it prefers full sun. As the plants often exceed 3 feet in diameter, they should be grown at least that far apart. Sage leaves should be harvested prior to blooming and dried in a well-ventilated room on screens or in a commercial dryer, away from direct sunlight and then store in airtight containers. Sow in plugs or seed flats 12 to 14 weeks before sale. Seeds to finished plugs, 8 weeks; plugs to saleable plants, 4 to 6 weeks.

Tarragon

A perennial herb the leaves of which are used for seasoning, especially vinegar. Tarragon grows to two or three feet tall and likes moderate sun, preferring some shade during the hottest part of the day. Tarragon, during growth, seems to have little aroma;

yet after the leaves or tops are harvested, the oils concentrate and start emitting their unique tarragon sweet smell. Plugs to saleable plants, 7 weeks.

Thyme

A plant of the mint family long cultivated and valued as a sweet herb. It has small lavender or pink flowers and is grown as a border plant, for ornament, or as an herb to be used for seasoning. Thyme should be planted in early spring. It is very hardy and will grow under most conditions. It prefers full sun. Thyme requires minimal fertilization when grown in a hydroponic system. Sow in plugs 12 to 14 weeks before sale. Seeds to finished plugs, 6 to 8 weeks; plugs to saleable plants, 4 to 6 weeks.

Watercress

Low growing and trailing European perennial, a member of the mustard family. It is easily grown from seed. Its natural season is from mid-autumn until spring. After its flower buds appear the leaves become too rank in flavor to be edible. It is also easily grown indoors in a hydroponic system. Start plants with seed by sowing lightly in pots filled with a medium. Watercress has many culinary, decorative, and medicinal uses.

Vegetables

Vegetables that do well in a hydroponic garden include artichokes, beans, lettuce, spinach, cabbage, beets, asparagus, broccoli, cauliflower, Brussels sprouts and peas. Vegetables that grow beneath the soil, such as onions, leeks, carrots, parsnips, potatoes, yams and radishes will also grow hydroponically, but may require extra care. Some crops to avoid are corn, zucchini, summer squash, and vining plants. They can be grown in a hydroponic garden, but they are not space efficient, and just not practical. They will dominate your whole unit. Your resources are better spent on crops more suited to the compact systems.

Warm weather crops: Start seedlings in March-April, transplant in May. Harvested by late August:

- Tomatoes.
- Green peppers.
- Eggplants.
- Cucumbers.
- Green beans.
- Squash (needs room).
- Melons (need room).

Cool season crops: Two growing seasons, early spring and early fall:

- Broccoli and cauliflower.
- Cabbage and bok choy.
- Lettuces.
- Spinach.
- Swiss chard.
- Peas, snow peas and sugar snaps.
- Green onions.
- Carrots (stubbies).

Beans

Beans are a warm weather vegetable that grows best in a Hydroponic system. Seeds are planted in the growing medium and watered from below by the bubbling system. Seeds germinate in 16 days, and since green beans are similar to other vegetable crops, no special formulation is required. They will grow fast, plump, and delicious.

Broccoli

Broccoli (from the Italian plural of broccolo, referring to "the flowering top of a cabbage") is a plant of the mustard/cabbage family. Broccoli has large flower heads, usually green in color, arranged in a tree-like fashion on branches sprouting from a thick, edible stalk. The mass of flower heads is surrounded by leaves. Many varieties of broccoli are perennial. It is planted in May to be harvested during the winter or early the following year in temperate climates. It grows well in hydroponic systems but since it is a heavy plant, needs sturdy staking.

Cabbage

Purchase cabbage seedlings or grow your own in rockwool cubes. When they grow large enough to stand on their own, transfer them to the hydroponic pots and surround the cubes with a growing medium. If you have a system that uses "flushing"–fill the tray from beneath with a flood of water, then drain it immediately–wait until your seedlings have grown large enough that they will not float out of the medium. Cabbage may need as much as two feet to spread out, depending on variety.

Cauliflower

Cauliflower, a cruciferous vegetable, is in the same plant family as broccoli, kale, cabbage, and collards. It has a compact head (called a "curd"), with an average size of six inches in diameter. It is composed of undeveloped flower buds. The flowers are attached to a central stalk. A sturdy support is important to keep all but the roots out of the hydroponic solution. This can be accomplished by either using a layer of gravel through which the roots can reach the hydroponic nutrients, or setting up netting that allows the roots to grow through. The hydroponic nutrients can be delivered through drip irrigation twice a day. This will keep the roots from getting too moist. Broccoli and cauliflower are two of the most cost-efficient vegetables to grow because of the lower temperatures required and the lack of need to continuously provide hydroponic nutrients.

Cucumber

Their quick growth in relation to the time available makes cucumbers a good candidate for a fall crop. They are normally seeded in July, set in the greenhouse during the first week of August, and harvested from September to December. The cucumber responds like a semitropical plant. It grows best under conditions of high temperature, humidity, and light intensity and with an uninterrupted supply of water and nutrients. Under favorable and stable environmental and nutritional conditions and when pests are under control, the plants grow rapidly and produce heavily.

Eggplant

Eggplant should be planted in full sun. The plants are easily injured by frost and will not do well with long periods of cool weather. Eggplants should be treated like tomatoes, the only difference being that eggplants like it warmer. Plant them from nursery stock, or starts, in the early spring. Allow 60-100 days to reach maturity from transplanting.

Lettuce

Leaf lettuce or semi-headed lettuce are the best varieties for hydroponic systems. Boston, black-seeded simpson, bibb or butter, or leaf lettuces are good choices. Lettuce requires additional light to grow successfully. It prefers cooler temperatures than warm weather crops. For best results, set your temperature controls for 75 degrees Fahrenheit during the daytime and 55 degrees Fahrenheit at night.

Peas

Snow peas like the cooler weather in the spring and fall, but don't handle the hot summer well. The same can be said for spinach, celery and most salad greens although they are slightly more tolerant. Chinese peas, also called snow peas are a cool season crop and are sensitive to heat. Temperatures over 86 deg. F will cause poor pod set. Before planting, you will want to choose the peas you want to plant: English (garden), Chinese (snow) or Snap peas. English peas only produce edible seeds. Chinese peas produce edible pods as well as edible seeds, and snap peas do the same. The peas will be ready to be picked around three weeks after the flowers start to appear.

Peppers

Peppers, like tomatoes, prefer warm growing conditions. Most peppers plants also require staking. Suggested varieties to grow hydroponically include cubico, mazurka, Fellini, Narobi and gold flame for sweet peppers and jalapeno, habanero and cayenne for hot peppers. Light and temperature requirements to grow peppers hydroponically remain the same as growing hydroponic tomatoes.

Spinach

Start your spinach seeds in rockwool. Place seeds in drilled holes and then hang slightly above a hydroponic nutrient solution in and ebb and flow system. Germinating plants need to be raised slightly to allow more oxygen to reach their roots, as the rockwool is not a good conductor in this aspect. After the seeds have sprouted, move them to their permanent spot. Allow twenty square inches per plant. The hydroponic solution you use at this point is extremely important as no growing medium is needed with the ebb and flow system. Plants are placed on a netting-type material with their roots allowed to dangle in the solution. Spinach will not thrive in too much light.

Tomatoes

You will achieve more success if you start hydroponic tomatoes from transplants rather than seeds. Try indeterminate, disease resistant varieties such as Trust or Daniela. Good cocktail or cherry tomato varieties for hydroponic growing are Cherita and Sweet 100s. Productive heirloom cultivars include Moskvich and Thessaloniki. Hydroponic tomatoes require staking or other support systems to maintain the plant in an upright position. Tomatoes need flowing nutrient systems where the feeding solution moves past the roots on a schedule four to six times per day. For seedlings, set the temperatures between, 68 degrees and 72 degrees Fahrenheit suggests the University of Arizona. Gradually increase the daytime temperature 70-79 degrees Fahrenheit and decrease the nighttime temperature to 61-65 degrees Fahrenheit. Use supplemental lighting up to 18 hours per day to improve production. Pollinate through vibration or gently shaking stems with blossoms.

Fruit

Fruit that can be grown hydroponically includes tomatoes, watermelon, cantaloupe, grapes, raspberries, blackberries, blueberries and strawberries. If you have the resources, some fruit trees can also be grown hydroponically, such as lemons or bananas.

Strawberries

Strawberry crops yield and produce well hydroponically. Growing strawberries hydroponically at home can produce enough berries to feed a family of four for a full year.

Production doesn't need to stop seasonally as a hydroponic system can be set up outdoors or indoors with artificial light.

A hydroponic system can be tailored to suit almost any growing application; strawberries are well suited for hydroponic cultivation. Perfect water and nutrient levels can be easily maintained to produce plump, juicy, unblemished fruit.

Always use runners that are certified virus tested. Carefully remove the runners or new strawberry plant from its container. Remove as much of the earth as you can by gently shaking and massaging the roots. Submerge the entire root system in a bucket of cold water for about 10 minutes then rinse roots under cold running water to remove any remaining dirt. Be very careful to inflict as little damage as possible to the roots and tiny root hairs as too much breakage will seriously stress plants and impair growth. Dry or brittle leaves and roots should be removed at this time. Add enough of the growing medium to fill the pot making sure the crown of the strawberry is well above the surface. The crown requires light and fresh air as this is where new leaves and flowers grow. If submerged, the crown will rot causing the entire plant to die. Insert the pot into your hydroponic garden. Once plants have finished fruiting and produced runners, clip the runners from the mother plant and root using conventional rooting methods. Once cuttings or runners have established a good root system, they must endure a chilling process. This may be done by placing rooted cuttings in a garage or cold cellar where temperatures remain between zero and 5C. Chilling may last anywhere from 10 days to five months depending on when your next crop is desired.

Exotic Hydroponic Plants

Indoor gardening offers some amazing opportunities when it comes to growing exotic, expensive and flavorful herbs and spices. We can create a tropical oasis where spices such as ginger, lemongrass, lime leaf, turmeric, chilies, curry leaf and even vanilla can thrive while delicate saffron is harvested from flowering corms and dried for year-round use.

There are many exciting species of spices to experiment with. Some of the lesser-known herbs such as epazote, used in Mexican cuisine, and Japanese perilla, also known as the sushi herb, produce well in hydroponic systems and add that extra special exotic touch to many dishes.

Hydroponic systems for exotic herbs and spices don't need to be complicated and fall into two main categories. For longer-term plants and those that produce rhizomes such as ginger, galangal, turmeric and kaffir lime, use deep media beds or shallow, free-draining beds for saffron corms. For shorter-lived, smaller plants such as perilla, epazote and many other herbs and spices such as dwarf chilies, use other systems ranging from nutrient film technique and pond or raft system to drip, ebb and flow or aeroponics.

Since most of the exotic spices and herb plants that can be grown hydroponically originate in tropical and subtropical climates, they are generally not fussy about day length, but they do like heat. Most ginger species, vanilla orchids, cardamom, turmeric, curry leaf plant, lemongrass and chilies like year-round warmth with an optimal temperature range of 72 to 860 F. Many of these can be grown outdoors in summer and then brought inside to overwinter in containers.

Spicy Species

For those who like Asian-inspired dishes, there are a good range of exotic herb and spice plants that mesh well together-kaffir lime leaves and fruit, fresh ginger, sliced galangal and lemongrass can all be grown hydroponically in a relatively small space provided they are kept trimmed and compact. For those who like hot, spicy curries, chilies, curry leaf and turmeric can also be grown in containers, although they need a little more space than the average herb garden. And for lovers of delicate dishes, saffron, produced by the crocus bulb, is one of the easiest spices to grow in a small hydroponic garden, with the advantage of also providing the grower with beautiful purple blooms.

Ginger Family

Traditional ginger rhizome (Zingiber officinale) is the most well-known crop in this category; there is also Myoga or Japanese ginger (Zingiber mioga), of which the young flower buds and leaves are used to flavor dishes. Another member is galangal or Thai ginger, which is similar in appearance to ginger but has a milder flavor. Galangal (Alpinia galangal) is a little more forgiving and seems to grow particularly well in hydroponics, being more cold tolerant in winter.

Ginger is an easy plant to start in hydroponics -fresh, healthy rhizomes can be purchased from stores or Asian supermarkets and sown into a moist hydroponic medium such as coconut fiber at 77 to 86°F. The knobbly buds on the rhizome will grow into fresh, new shoots with roots developing from the rhizome. If conditions are warm enough, ginger plants will then produce a large number of shoots and an expanding root system of thick fleshy rhizomes, taking anywhere from 12 to 18 months before the first harvest of large rhizomes can occur.

The plants are best mounded up as the rhizomes develop adding more growing medium around the base of the plant helps increase yields and quality of the harvested product. General-purpose nutrient solutions at an EC of 2.2 to 2.6 can be applied, but using a fruiting or bloom formulation at a higher EC seems to assist with intensifying the flavor profile of the rhizomes.

Another member of the Zingiberaceae family is turmeric (Curcuma longa), which, like ginger, grows from a rhizome that can often be purchased fresh in Asian specialty stores. Turmeric can be grown hydroponically in much the same way as ginger and used fresh or dried and ground to create the distinctive yellow, powdered spice.

Saffron

Saffron (Crocus sativus) has long been sought after as an expensive, delicate spice that adds both color and distinctive flavor to many dishes. Saffron is produced commercially as an outdoor field crop in countries such as Spain, India and Iran, where the climate is suitable and cheap labor is available for the time-consuming harvesting and processing of the flowers. However, saffron is an easy, small-scale crop for hydroponics and these days saffron bulbs are readily obtainable for those who want to grow it.

Saffron is the world's highest-priced spice and is composed of the bright yellow, dried stigma of the saffron flower. The saffron crocus grows from a small, rounded corm(similar to a bulb), which after flowering and vegetative growth, multiplies by the production of many small daughter corms. Each flower produces on average three strands of saffron, weighing in at much less than a gram, hence a huge number of flowers must be grown, picked, processed and dried to make just 100 grams of this spice. For the hobby hydroponic grower or household chef, a small plot of saffron should give sufficient spice to last for many months as only a few strands are used in most dishes.

While it is fairly easy to grow, saffron does need some specific requirements if the objective is to propagate and multiply the plants. There are two ways of running a saffron system. The dormant corms can be purchased, usually towards the end of summer or in early autumn, then planted, flowered, harvested and the corms discarded, all over a six-week period. Or after the short flowering period, the bulbs can be grown on, producing vegetation and new daughter cormlets over a period of many months until they become dormant in mid-summer. The first system means the corms are planted at a high density since they won't be grown on and multiplied.

The second system needs more space for the plants to fully develop and many months of caring for the corms after they flower, although the reward is a supply of new corms that can be sold or given away to others and a higher yield of flowers in the next year. Tying up your hydroponic system with vegetative saffron for nine to 10 months may not be attractive to those with limited space, although trays of saffron plants are cold hardy and can be put outside for winter and spring if necessary.

Corms can be stored in a dry place and planted out when the hydroponic system is ready. The spacing should be approximately 4 to 5 in. apart and 1.5 to 2 in. deep, in a tray of free-draining, sterile growing media such as coconut fiber with some perlite mixed in. The media needs to support the plants, but at the same time be friable enough for the young corms to form without deformities, so any substrate used for baby root crops would be suitable.

The flowers will have already been initiated in the corms during the summer dormancyperiod and flowering will occur when moisture is provided and temperatures start to drop in fall. Flowering is triggered by environmental conditions such as temperate and moisture, which is easily manipulated in a hydroponic growroom. The ideal conditions during flowering are: a 16-hour day length with day temperatures of 63oF and night temperatures of 54o F. An indoor growroom or greenhouse situation means the flowers are protected from rain, moisture and wind, and the lack of weeds makes harvesting high-quality flowers much easier.

Flowering will typically occur quite quickly after planting and within a few weeks, the first emerging flower buds should be seen. The flowers will fully open in three to five days and be ready for harvest. As each flower blooms, it should be plucked or snipped from the plant and taken away for processing. The easiest way to remove the saffron stigmas from the center of the flower is to pull back and remove all the petals and then snip the golden strands at the base. The strands will need to be dried before storage.

Citrus Flavors

Citrusy, aromatic kaffir lime leaves (Citrus hystrix) and fragrant lemongrass (Cymbopogen citratus) are distinctive and popular flavors associated with many Asian and ethnic dishes and also easy to grow hydroponically. Kaffir lime (also called Makrut lime) is a small, compact tree grown mostly for its distinctly flavored leaves, but the peel of the small knobbly fruit is also ground into a paste and used in many Asian dishes and curries.

Kaffir lime trees can handle cooler conditions than fruiting lime types and will produce for many years. In an indoor garden, they are best kept regularly trimmed by harvesting foliage for culinary use so the tree remains compact and productive. Ideal temperatures for growth are 72 to 90 °F, and the plant should be grown in a large pot or grow bed of media such as perlite, coco fiber or rockwool with a drip irrigation system. Warmth, high light, increased EC (2.2 to 3.0) and slight moisture stress will concentrate flavor and aromatic compounds in the plant, resulting in higher-quality harvested spices.

Lemongrass is a popular hydroponic herb, grown commercially for fresh harvests. Lemongrass seeds germinate rapidly within three to five days at 77 to 86 °F, making this an easy plant to establish. Mature plants can become as large as 4 ft. in diameter with leaves 3-ft. long, but small plants can be grown if kept well-trimmed.

Even the foliage of young seedlings contains the fresh pungent flavor of lemongrass, so harvesting can take place as soon as there is sufficient foliage to cut for use. The use of small rooting volumes such as nutrient film technique is not recommended due to the large and fibrous root system that rapidly develops and can cause system blockages and slow the flow of nutrients.

Lemongrass requires warm temperatures to produce well and develop a strong and distinctive flavor, although it will survive cool conditions in a dormant state. Optimal temperatures are 72 to 95 °F, with a high humidity level (more than 80%) and full sunlight or high levels of artificial light. Full-spectrum grow lamps are recommended for many spices such as lemongrass as there is evidence to suggest certain parts of the light spectrum such as UV assist with the development of flavor compounds in plants.

Epazote and Perilla

International cuisine has introduced us to some exciting exotic herbs, many of which have now become hydroponic crops. Two of these are epazote, often considered to be an essential ingredient in many traditional Mexican chili sauces and bean dishes, and Japanese perilla or shiso, also known as the sushi herb. Both are small annual plants, easy to propagate from seed (perilla seed should be sown while still fresh) and grow well in nutrient film technique or media-based systems under similar conditions to those used for basil and cilantro.

Perilla (Perilla frutescens) is an attractive herb that is available in both deep purple and green forms with a flavor similar to mild basil. Perilla grows best under warm conditions of 64 to 79 °F with full light and takes around 35 to 40 days to mature. It is also grown as a microgreen species harvested after the development of the first seedling leaves. Epazote (Chenopodium ambrosioides) is a small annual plant with a distinctive flavor often described as camphor or turpentine-like with a strong aroma. Easily grown from seed, epazote is a rapid growing herb requiring temperatures of 64 to 79 °F, full light and with similar nutritional requirements as other hydroponic herbs such as basil.

A hydroponic garden full of exotic herbs and spices originating from faraway lands, ready to pluck and serve in fresh, fragrant dishes is a real possibility with a climate-controlled indoor garden. Many of these species are hardy plants, easy to grow and thrive under the advanced nutrition hydroponic systems provide. Even the smallest growing area can accommodate a saffron corm or two, making for eye-catching displays as well as impressive harvests.

Fast Growing Hydroponic Plants

There is no need to have a backyard or garden for fast growing hydroponic plants. You only need a small space and the right equipments to set up. Others use a wide greenhouse or a storage room depending on the volume of plants of to grow. This kind of method is more efficient and productive than the regular one as you're able to plant as 4 times more.

Another advantage of growing plants through hydroponic is that it prevents pests, insects and diseases that are caused by soil. It is definitely a much safer and more organic way of growing plants and harvesting produce. Furthermore, there is less waste or manure and fertilizer to use during the whole process.

Best Choices of Fast Growing Hydroponic Plants

Lettuce

Lettuce is one of most popular leafy greens consumed at homes and in restaurants because of its versatility. You can add to your burger, salad and even a smoothie glass. In planting, this is also one of the best and fast growing hydroponic plants. You may choose to use any variety from Romaine to Boston, bibb to butter crunch, crisphead to looseleaf. Most of these are ready to harvest in three weeks to a month. When you do this, make sure to pick only the outer leaves and leave the inner parts so that you can harvest them the next day.

Spinach

Spinach is also a great choice to plant as this is among the fast growing hydroponic plants. Like lettuce, it is also used in many dishes since it's not only crunchy and delicious but also rich in nutrients. This is an excellent source of vitamins A, C, E, K, copper, folic acid, iron, magnesium, protein, zinc and so much more.

There are other leafy greens that you can grow in your hydroponic garden, such as kale, watercress, arugula, Swiss chard and mustard greens.

Cucumber

Cucumber is full of water, about 90%, which makes it a great choice to grow through hydroponic system as well. This fruit has several uses aside from being a part of a meal. It can be used to treat some skin issues and eye bags as well.

Tomatoes

It is very easy to grow tomatoes in soil but this can be also sown through hydroponic method. You can definitely make your own supply of tomatoes usable as meals, juice and paste. It is rich in phytonutrients, most notably lycopene.

Peppers

Included in this list of fast growing hydroponic plants are red, yellow and green peppers. This vegetable is loaded with nutrients. When growing it through this system, it is best to decrease the temperature at daytime, while increase at night once the plant has reached its maturity.

Spring Onions

Spring onions are not just a type of spice but a wonderful plant eaten because of its richness of vitamins and minerals. This is very easy to grow through hydroponic system. You can quickly harvest some before they become fully mature.

Basil

Most herbs are fast growing hydroponic plants and basil is one of them. This plant is a good source of anti-inflammatory properties and antioxidants. In short, it is a healthy plant that can help prevent certain diseases and slow down aging.

Coriander

Coriander is another herb that can be grown through hydroponic method. It takes only a month to pick some that you can add as seasoning to your dishes. Like most herbs, it has great benefits to the health.

You may choose other herbs, such as peppermint, spearmint, oregano, chives, thyme and marjoram, to grow in a hydroponic garden. Any kind is easy to plant since they grow much faster, 25% more than the traditional gardening method.

These are only a few of the fast growing hydroponic plants you can try for your new garden.

Mint

Mint is undoubtedly best crops for hydroponics it grows like a weed with freshness guaranteed nearly anywhere it is planted. There are a diversity of types of mint, but the mainly used varieties are spearmint (Mentha spicata), peppermint (Mentha x Piperita), and pennyroyal mint (Mentha pulegium); some of the other mints like lemon mint (Monarda citriodora). Mint display leaves of various colors you can find mint from grey-green to dark green, green and purple, to sometimes blue and pale yellow too.

With mint, you can grow more crops, in less space, and with less environmental restriction. Undoubtedly mint is one of the most vigorously growing plants in any medium. The aromatic compounds present in mint are pungent, refreshing, and exceptional. But they also lose their freshness pretty rapidly, which means that mint has a limited shelf life. Once you've had freshly harvested mint, there's no going back either you finish it or leave it to spoil.

The uses of mint are nearly endless. Everyone has heard of mint, whether in teas, jellies, desserts, garnishes, or as a seasoning for savory dishes and ladies savior as granny's beauty secret. Moreover, it is a key element for drinks like mint or mojito, where mint's flavor is key to a great drink and blast of freshness. Mint is an aromatic, fast-growing herb that adds a refreshing flavor to any dish you add it to.

A mint plant (Pudina).

Mint is one of the most recognized and popular hydroponic herbs to grow in a hydroponics system. With its characteristic aroma, fresh and sweet taste, mint is difficult to resist. Flowers on mint are usually shades of white to purple. Mint is not a plant you frequently have to worry about propagating it does a great job on its own, spreading out its stoloniferous roots in all directions to colonize the entire space if you plant it without a root barrier.

Considerations for growing hydroponic mint include appropriate light, nutrients, temperature, humidity, and pH level. Once these factors are controlled, hydroponics offers a less time-consuming and superior growing medium in comparison to the conventional soil or best answer to how to grow mint indoors.

- Mint can also be started from seed but is more commonly begun with cuttings/clones or rootstock. For propagating mint simply pre-soak all growing media mainly Rockwool and vermiculite in a pH balanced water for at about 45 minutes before using it so that it provides sufficient moisture for growth, on the other hand, dry media will soak up moisture from the plant's roots resulting in stress. Moistened propagation cubes work well for seedlings, similarly as does a conventional soil mix. Germination usually takes about1-3 weeks. Once the seeds are displaying their true leaves, the plants can be transplanted to your hydroponic growing system. It must be noted that seed germination is little difficult and uncommon when growing mint a plant from a cutting (or an established plant) is found to be more successful and vigorous than propagating from seed.

- Growing mint needs about 12-16 hours of bright, unobstructed sunlight per day. Keeping your hydroponic setup in the south-facing window will provide adequate lighting to the herb. Keep rotating the plant to ensure all sides receive sufficient light. Since grow lights are recommended for extra supplementation, especially cases where your plants are not receiving the minimum natural sunlight to stay healthy. Standard fluorescent lamps may be enough, but their functioning is less effective in comparison to LED grow lights. High-intensity discharge lights (HID) are also found effective particularly metal halide, which is proven superior for growing leafy herbs but it is expensive too. Since grow lights with blue spectrum lighting encourage lush, bushy growth in mint so grow lights with blue spectrum potential may be the best option to grow your hydroponic mint.
- Humidity is maintained as 70 – 75% during the rooting phase, if growing from cuttings humidity should be around 85 – 90% until roots are established which a phase of 7 to 10 days is generally.
- Temperatures: During night cycle should be 50 – 55 Fahrenheit, Day Cycle: 65 – 70 Fahrenheit. Make a notice that temperatures above 85 Fahrenheit affect and may halt mint growth.
- Ideal pH range: 5.5-6.0.
- PPM: 1400-1680.

Hydroponic Mint Nutrient Solution

Growing mint in the hydroponic jar.

It is specifically formulated for the hydroponic system are another important component of growing hydroponic mint like all the desired nutrients or say food for development is present in it. Combining required nutrients and water creates the "nutrient solution". It is suggested to drain out, clean and refill a new batch of nutrient solution regularly to maintain optimal growing conditions. This process should be done at least once a month, although more frequently is encouraged. Mint doesn't request a lot so far nutrients are concerned; a simple balanced formulation available in various stores is all that is needed. Do not use a high nitrogen formula as it may promote lush green mint leaves that are grossly substandard in flavor. Various nutrient solutions brand has differing concentration and water to nutrient ratios. It is better to follow the directions that came with your nutrient solution.

Hydroponic System for Mint Planting

The hydroponic system is nothing but a framework which facilitates the nutrient solution delivery to the developing plants. In case of hydroponic mint methods for providing the nutrient solution take account of the Passive method (which allows you to decide when and how much nutrient solution a plant needs, without using pumps and timers), the Flood and Drain method (when trays and pots are flooded with the nutrient solution using a water pump), a Drip System (method which uses a pump to drip nutrient solution on to the plants), the Nutrient Film Technique (NFT), aeroponics and the mint aquaponics system.

Growing Media for Mint in Hydroponics

we have talked about the nutrient solution, setup, pH but what holds the suspended plant in the hydroponics system yes it is the growing media every crop has its set of growing media. In the case of hydroponic mint, there are a variety of mediums to choose from. Some of the most popular comprise of coconut coir, lightweight expanded clay aggregate (L.E.C.A), perlite and vermiculite. These mediums are inert and act as an anchor to the plant but do not offer the essential nutrients for the plant's health. Thus, the growing medium you choose is of lesser significance than that of maintaining a continuous delivery of the nutrient solution.

Some Facts of Growing Mint

Facts of growing mint.

- Mint can be grown indoors in pots or containers with soil or without soil (Hydroponics).
- Growing mint hydroponically or in water is very easy.
- Mint is a perennial herb and once planted, you can have multiple harvests.

- Mint is a fast-growing herb and usually grows 3 to 4 inches per month or can spread up to 2 feet in 5 to 6 months.
- Mint can be grown in partial shade as well.
- A well-balanced all-purpose liquid fertilizer is beneficial in soil-grown mint.
- The mint plant good at repelling bugs away.
- Mint plants require a lot of water when it is grown in the soil under full sun.
- To protect mint plants from various pests and bugs, you can use organically prepared pest controls such as neem oil, garlic spray, and soap-based solution.

Hydroponic Ginger and Turmeric Production

Ginger and turmeric are members of the Zingiberaceae family of tropical and subtropical plants. They are well-suited to protected cultivation in a warm, humid environment with plentiful nutrients and moisture-retentive substrate beds.

Fresh rhizomes of ginger and turmeric have long been recognized for their culinary properties. They provide a slightly different, more aromatic flavor profile than their dried counterparts, owing to the presence of volatile compounds that are released when the rhizomes are cut or grated.

Aside from their use in edible dishes, ginger and turmeric both have several medicinal uses, with fresh ginger used to alleviate nausea, while turmeric has long been used for its digestive, antioxidant, and anti-inflammatory effects. Turmeric contains many powerful compounds, including different curcuminoids and volatile oils like zingiberone, turmerone, and atlantone. Ginger contains gingerol, a natural bioactive compound known to have anti-inflammatory and antioxidant properties, as well as be useful as a digestive aid and as an antibacterial agent.

Along with these medicinal and culinary properties, a well-developed ginger or turmeric plant provides a lush, tropical feel to any indoor garden, with their large, lime-green, attractive leaves. Turmeric plants also have the advantage of producing beautiful exotic blooms if conditions are conducive to rapid growth and development.

Growing ginger and turmeric hydroponically ensures the rhizomes can be harvested at any stage. Fresh, young ginger rhizome has a crisp and juicy texture that lacks the fibers of older rhizomes. Its flavor is subtle and ideal for slicing thinly to eat raw or to make into pickled ginger for use in Japanese cuisine. Fresh turmeric has a mild, earthy flavor, and a finger-staining, intense yellow color that is often used as a natural dye. Young harvested turmeric rhizome can be used immediately, grated into curries and other dishes, or blended with fruit and vegetable juices to create a supercharged smoothie. The harvested rhizomes may also be air dried and will keep for many months before replanting for the next crop or being consumed.

Propagation

Ginger and turmeric are surprisingly easy to propagate from rhizomes purchased from the supermarket, specialist Asian grocery market, or online. Both are sold as fresh rhizomes, although they may have been in storage for several weeks to months before sale. In this dormant state, the buds present on the sides of the rhizomes remain relatively flat and difficult to see. Once conditions become sufficiently warm, these buds start to swell. Small buds forming on ginger rhizomes kept in a warm kitchen after purchase are common, but turmeric often needs to be sprouted under warm and moist conditions before the young sprouts will form and become large enough to see.

Turmeric contains plenty of buds on the sides and ends of the rhizomes, and even old and dried up rhizomes can spring back to life once planted out into moist substrates. After obtaining fresh rhizomes, they can be broken into fingers or side shoots to increase the number of plants obtained from each even small sections of rhizome usually contain at least one bud that will form a new shoot and young roots once planted.

Optimal growing mixes during the propagation stage are sterilized perlite, vermiculite, coconut fiber or rockwool, which helps retain moisture but at the same time won't become oversaturated, which can encourage rhizome root. Pieces of rhizome are pushed into the growing substrate in a shallow tray to a depth of around two inches with any visible buds facing upwards. The tray is then best placed on a heated propagation mat or pad to speed up the rate of shoot formation. Ideal temperatures surrounding the rhizome pieces are around 82-90°F in the growing media at this stage. Buds should start to swell and grow along the sides of the rhizomes within two to four weeks, and as they emerge, young roots will form around the base of the shoot.

First Shoot

After the first leaf has emerged and unfurled, the sprouted rhizome can be transplanted to a larger container or growing bed and the first application of dilute nutrient solution applied. As the rhizomes are developing the first shoot, a moderate level of light that is typically used in a propagation area for clones and cuttings can be used, with a relative humidity level of 80 to 90 per cent. Because the plants produce an underground crop of rhizomes, a soft, friable, moisture retentive growing substrate is ideal. Coco fiber fines mixed with 20 per cent perlite or a vermiculite/perlite mix is ideal, although the plants are quite adaptable to a range of other substrates. Large pots, buckets, and beds are ideal, provided the depth of substrate is at least 12 inches to allow for the large size of the root system at plant maturity.

Systems and Growing Conditions

Drip hydroponic systems are most suitable for growing ginger and turmeric, and these heat-loving plants will happily grow alongside tomatoes, capsicums, cucumbers, melons, and other fruiting plants because they have similar nutritional requirements. Once ginger and turmeric plants have two to three leaves, light levels can be increased to full strength, with similar intensities as many other high light crops grown in indoor gardens. While shading can be tolerated and doesn't detract from the attractive nature of the foliage, it does reduce rhizome yields and increases the time to harvest. Ginger plants are best mounded up as the rhizomes develop. Mounding refers to adding more growing medium around the base of the plant, which helps increase yields and quality of the harvested product.

Day length is not an issue with these plants, but they do need warmth with optimal levels of 72-86 °F, although both will happily grow at temperatures higher than this. Cool conditions will slow or prevent growth, so these plants are ideal candidates for growing outdoors in summer and bringing inside to overwinter and continue producing in winter in the tropical heat and light of a well-lit indoor garden.

Humidity

Ideally, ginger and turmeric prefer humid conditions of 70 to 90 per cent. By growing them in large, densely planted clumps, they can create their own micro climate of humidity, and this, combined with a damp growing medium, tends to lead to optimal growth. Low humidity will cause the lower, older leaves to develop slight tip burn, which can be trimmed if necessary. Placing ginger or turmeric plants in the most humid area of an indoor garden is the best course of action.

Nutrients

General-purpose nutrient solutions at an EC of 2.2-2.6 can be applied to ginger and turmeric crops. However, switching to a fruiting or bloom formulation at a higher EC seems to assist with intensifying the flavor and aromatic profile of the rhizomes. Your pH levels are best maintained around 5.7-5.8 to maximize nutrient uptake.

Reaching Maturity

While both plants are long-term crops, turmeric will produce harvestable rhizomes more rapidly than ginger plants. Under ideal growing conditions, with sufficient heat, the first small turmeric rhizomes grown hydroponcially can be harvested in as little as six months, with main crop yields taking around eight to nine months. Ginger can also be harvested relatively young and small; however, full-sized, large rhizomes can take anywhere from 12 to 18 months from planting.

Ginger plants at maturity are generally larger than turmeric, although this depends somewhat on variety and growing conditions. Ginger plants should be spaced 18 inches apart, and turmeric plants 12 to 14 inches apart, provided light levels are sufficiently high. This will give a dense canopy of foliage that makes a tall backdrop for smaller ornamental foliage and flowering plants in amenity plantings. While the flowers of culinary ginger are insignificant and not particularly attractive, turmeric blooms are delightfully exotic, won't affect rhizome development, and are even edible, so these alone are worth growing the plants for.

Harvesting ginger and turmeric is relatively simple. Turmeric plants will develop foliage yellowing and dieback once the rhizomes are fully mature. Plants can be harvested before this stage as young rhizomes are tenderer and fully usable, but they won't store for extended periods of time. Plants can be pulled from the growing substrate, which has been allowed to dry down for a few days, and the rhizomes plucked from the root system.

Alternatively, a rhizome or two can be generally removed form the plant while it is still growing by digging down around the base of the stem and removing the largest that has formed, then replacing the substrate. The plant will continue to direct reserves into the remaining rhizomes, allowing a longer period of fresh rhizome harvest. Once harvested, the skin on ginger and turmeric rhizomes is so thin it can be gently washed before immediate use. For longer term storage, they require air drying to cure the outside layers of the rhizome or stored wrapped in plastic under refrigeration.

Pest Control

Under optimal growing conditions, ginger and turmeric are relatively hassle-free crops and seem remarkably immune to common exotic plant problems such as root rot or die back due to overwatering. They can, however, suffer from common pest issues. With an indoor garden or greenhouse these typically include fungus gnats during the propagation phase, and mites, thrips, and whiteflies on more mature plants, which can all damage and discolor the foliage rapidly.

Early identification and control of pests is recommended, and the large, dense canopy may also provide a suitable environment for the use of integrated pest management (IPM) with introduced predator and parasite insect controls. The foliage of turmeric can be prone to spray damage from the regular use of insecticide oils, so these are best avoided unless plants are close to harvest maturity when leaf marking is less of an issue.

While ginger and turmeric may be familiar as dried spices readily available in pre-packaged form, the young rhizomes freshly harvested from a hydroponic garden provide a more powerful punch when it comes to flavor and bioactive compounds.

Oregano

Oregano is a small, bushy herb with a strong unique flavor that's especially pungent when the herb is fresh. The leaves are used fresh and dried in most types of cuisine, but especially Italian and French.

There are three main kinds of oregano used for culinary purposes: Greek (Origanum vulgare hirtum), Mexican (Lippia graveolens, which actually isn't oregano at all), and Italian (Origanum x majoricum). The phenol responsible for the oregano taste (carvacrol) is also found in other species.

Oregano is sometimes confused with marjoram, which is a similar but separate herb. Italian oregano is a hybrid of marjoram, however.

Each oregano has slightly a different flavor which varies even more based on conditions, but all are similar in their preferences.

Oregano has small, rounded leaves that are fuzzy in some species this makes it harder for them to deal with high humidity.

You can buy oregano in bulk for $60–65/lb, which ends up being about $4/oz. Dried oregano is $2.50 4.00/oz and dried organic oregano comes in higher at $5–5.50/oz. Fresh oregano pricing varies by market and grower, but Upstart Farmers can consistently receive at least $2–3/oz for herbs.

Because oregano is a popular herb, demand isn't difficult to secure. Be sure to check in with your local markets, however; your crop choices should heavily depend on real-life feedback that is specific to your situation.

Oregano

Origanum spp. and others:

- Most common pests and diseases: Thrips, whiteflies, spider mites, etc.
- EC/PPM: 1.5-2.0.
- pH range: 6.0-8.0.
- Temperature: 55–70°F.
- Yield: 2.5 lbs/tower/8 week turn.
- Planting: Propogation by cutting is best.
- Harvesting: Multiple harvest; prune similar to rosemary.

Oregano can bring in a lot of profit, but be aware that the growing timeline is much longer. Oregano is a slow grower the first harvest is 8 weeks after transplant. Once established, the tips can be harvested every 3 or 4 weeks.

In difficulty, oregano ranks with other woody herbs like basil, rosemary, and thyme. It takes a bit more care than one-harvest crops like lettuce, needs dry periods in irrigation, and thoughtful pruning. Personally, I think of oregano like the caviar of herbs; it's incredibly tasty, often gets higher pricing, and it's worth the work.

Once harvested, fresh oregano sprigs have a shelf-life of 1–3 weeks. Be careful of over-chilling or excessive moisture on the foliage. Upstart Farmers like Haydn Christensen at Bayberry Fresh are packaging their oregano in clamshells for sale at grocery stores and farmer's markets.

Ideal conditions for oregano:

- pH: 6.0–8.0.

- EC: 1.5–2.0.
- Temperature: 55–7 °F.

Oregano suffers from typical greenhouse pests like thrips, whiteflies, and sometimes aphids. Keep your IPM strategy up to date and keep an eye out for signs like brown spots, yellowing, and wilting. Be especially aware of overwatering: like rosemary, oregano prefers low watering and is susceptible to root rot and other diseases if kept too wet.

Oregano can be started from seed, but you can both shorten timelines and achieve more consistent tastes if you propagate oregano from cuttings instead. Use a rooting hormone (you can buy this at any store with garden supplies) and a substrate with good water-holding capacity; rockwool or peat work great. Cut sprigs from an oregano plant, wet the ends, and dip them into the hormone powder. Wrap the dusted sprig in the substrate. When 5–10 roots are sprouting from the sprig, it's ready for transplant. This could take up to 6 weeks.

Once transplanted, oregano typically takes about 8 weeks to mature and be ready to harvest. After that, it can be harvested in 3–4 week turns. Harvest the tips of the branches during the cool part of the day and take only about a third of the plant, leaving the rest for it to photosynthesize and regrow.

Basil

Warm-season, aromatic herbs such as basil and cilantro thrive outdoors in the long, hot days of summer. However, greenhouses and indoor gardening have seen these popular crops become a year-round feature in many kitchens and restaurants.

What makes these herbs stand out from many others is the fact they lose much of their flavor and aroma when dried and thus tend to be used fresh. While it's relatively easy to grow basil and cilantro indoors, there are a few tips to keeping the flavor, aromatics, essential oils, and overall quality of these herbs high while producing plentiful supplies through the long winter months.

Indoor hydroponic systems can also take advantage of the diverse range of basil cultivars now available as seed, many of which are difficult to obtain as a cut herb product. Modern hybrids also offer an increased range of disease resistance, easy to handle pelleted seed, slow bolt types, and high-yielding varieties specially bred for container growing and protected cropping.

Basil Varieties

While most of us are familiar with the traditional, large-leaved sweet or Genovese basil used in a huge range of dishes, often associated with tomato, chillies, peppers, in hot salsas and blitzed into pesto, basil varieties are far more diverse than just these cultivars.

Not only are the glossy green leaves of basil highly attractive, but they are also available in a number of more colorful ornamental forms. Purple, frilled, ruffled and spotted, mottled and bi-colored cultivars are available that not only have that distinctive basil flavor, but brighten up hydroponic systems with a splash of intense color.

Plant size and form also differ between basil varieties, with some highly compact types having been bred for container production, while flavor profiles in basil are diverse, giving great opportunities for culinary use. Of the purple types, Red Rubin, Dark Opal, Purple Ruffles, Purple Delight and Basil Bicolor are well-suited to hydroponic production, with high light levels that are required for a deep, intensive coloration to develop.

Of the sweet green types, those suited to indoor gardens include Genovese Compact Improved and Elindra, bred for container/greenhouse production. These cultivars are great for those short on space. Nufar and Aroma 2 are taller Genovese sweet basil types that have the advantage of resistance to fusarium root rot disease.

For basil destined for pesto production, specific varieties include Lettuce Leaf Basil, Napoletano and Italian Large Leaf cultivars, which produce sweeter foliage than Genovese types with leaves up to 4 in long.

More exotic basil varieties are used in a range of cultures and cuisines and many are well-suited to hydroponic production. These include the popular:

- Sweet Thai Basil with its distinctive anise-clove flavour.
- Holy Basil (Kapoor Tulsi) with a spicy flavor and aroma.
- Cinnamon Basil.
- Citrus or Lemon Basil and Lime Basil with a strong citrus/basil flavor combination.
- Spicy Basil.
- Globe Basil with small leaves suited to restricted spaces.

Cultivars of basil with inbred disease resistance or tolerance to common diseases downy mildew are worth considering for larger-scale hydroponic basil production.

Basil Propagation

Basil is most commonly propagated from seeds, which are available in raw and pelleted form. However, cuttings can also be taken and will form roots within seven to 10 days.

Seed germinates readily under warm conditions with 75 °F being ideal for most basil types and takes between five to seven days. Rockwool propagation blocks are often used in commercial production, while perlite/vermiculite, coconut fiber, peat moss, foam substrates, and sand culture are also suitable provided the medium is sterilized.

Basil seedlings are extremely prone to pythium and other damping-off pathogens, so care needs to taken with watering in the early stages post germination to ensure that the surface of the growing media is not overly wet.

Once the seedling leaves have expanded, a dilute nutrient solution at an electrical conductivity (EC) of 0.5 $mScm^{-1}$ can be applied with a pH between 5.8-6.2. Once transplanted into a hydroponic system the EC can be gradually increased and full, high-intensity light levels supplied.

Tall, dense stems of many basil varieties benefit from a good degree of air movement and ventilation, up and under the plants to help prevent some of the fungal infections such as botrytis (grey mold), which these plants can develop under high humidity.

Just as with cilantro, basil can be prone to bolting at a fairly young age if the plants are stressed. This may include root restriction, moisture stress, overly dry environments, excessive light or temperatures and overcrowding.

Flower buds can be removed from the growing points of plants as they develop; however, once these are seen it is usually more economical to replace with young transplants unless blooms are to be harvested for culinary use.

Essential Nutrients for Basil

Many hydroponic growers use a general-purpose lettuce/herb or vegetative nutrient formulation for basil. While this is suitable for young plants, more mature basil often has a different nutrient uptake ratio.

The amounts of potassium (K) and calcium (Ca) are relatively high in the edible portion of the basil plant. For this reason, the ratio of K:Ca in the nutrient solution needs to be kept at a reasonably high level close to 1:1.

While potassium and calcium are directly rated to the oil levels and flavor in the foliage, nitrogen (N) is essential for leaf yields and needs to maintained at all times, while magnesium (Mg) should be at levels of at least 50 ppm as there is a direct relationship between Mg content, yield and composition of the essential oils that make up much of the characteristic basil flavor and aroma.

Harvesting your Basil Crop

Basil grown in an indoor garden may be harvested in a similar fashion to those grown commercially in a greenhouse as a cut and re-grow crop with only one-third to two-thirds of the upper foliage taken as stems, leaving the plant to regenerate for a further two to three cuts before replacing the plants.

As an alternative, the lower mature leaves can be individually plucked as required and the growing tips removed as soon as the plants reach a reasonable height to promote the development of more stems and to help keep plants compact. An oversupply of hydroponic basil can be used to create not only products such as pesto, but infused oils, vinegar and salsa. Some varieties of basil may be dried and used as teas while others have medicinal uses.

Freshly harvested basil has a limited shelf life as the volatile compounds which contribute to the distinctive flavor deteriorate rapidly. Basil has specific requirements once cut ready for use and should not be stored under refrigeration or below 50 °F as chilling damage will occur and leaves will blacken. Harvested basil should be lightly wrapped in plastic and stored at temperatures between 52 °F to 57 °F. Cut stems can be placed in a glass of water at room temperature to maintain shelf life for several days.

Cilantro

Cilantro (Coriandrum sativum), sometimes referred to as leaf coriander, Chinese or Mexican parsley or simply as coriander in many countries, is a fast-growing, warm-season annual herb.

Cilantro and coriander are the same plant cilantro refers to the flat, fan-shaped leaves, while coriander refers to the seeds that are ground and used as a spice. Cilantro leaves have a characteristically different aroma and flavor profile than the ground seeds, but both are frequently used in the same dish.

The cilantro flowers are also edible, used raw in salads and as a garnish. Cilantro is a fresh herb with a limited shelf, its flavor and aroma diminishes with shipping and storage and the delicate foliage is easily damaged.

For this reason, it is an excellent herb to grow in an indoor garden to be harvested just before use when its flavor quality is highest. Since cilantro has a habit of going to seed (bolting) fairly rapidly it is important to select cultivars that will not only grow large, quality foliage, but are also slow to bolt in the warm, protected environment of an indoor garden.

Varieties such as Calypso, Santo and Marino are well-suited to hydroponic production, as they are slow to bolt and allow for two to three harvests or cuts before plants need to be replaced.

Cilantro Propagation

Cilantro seed is large, rounded and easy to handle and is best sown into individual cubes, cells or pots of sterilized, free-draining substrate, as seedlings are not suited to transplanting. Germination occurs within five days at 68°F to 78°F, with seedlings requiring warmth (above 68°F) and moderate light levels to prevent elongation and tall, weak growth.

Under hydroponic production many growers sow several seeds into each planting site or pot and grow these mini bunches for just a few weeks before cutting. By using successional sowings every two to three weeks, fresh cilantro can be harvested year-round, minimizing the issue of plants bolting after two to three harvests.

Harvesting your Cilantro Crop

For small-scale indoor gardens, cilantro can be harvested at any stage from young seedlings (microgreens) through to mature plants, depending on the flavor intensity required. Younger plants are tenderer and the flavor milder, while older plants, particularly those about to flower, develop stronger, sometimes slightly bitter flavors.

Freshly cut cilantro has a high rate of respiration and should be used within a few hours of harvest. Where cilantro needs to be stored after cutting, refrigeration in plastic packaging at 32°F gives the longest shelf life and prevention of post-harvest wilting.

Protecting Cilantro from Garden Pests

Cilantro is particularly prone to infestations of aphids and whiteflies, which can become major pests if not controlled early. Mites may also be an issue under hot, dry conditions, while high humidity can predispose plants to botrytis and bacterial leaf spots.

Maintaining good airflow around the base of dense cilantro clumps assists with disease prevention, while sprays containing neem or neem oil are beneficial to break the life cycle of many insect pests of cilantro.

Hydroponic Systems for Basil and Cilantro

Both basil and cilantro can be grown in a range of hydroponic systems, the most common being nutrient film technique, DFT (deep flow technique), aeroponics and shallow substrate systems using coco coir or rockwool mediums. Float, raft, pond or raceway solution culture systems can also be used, provided the depth of the nutrient solution is at least 5 in.

Temperatures of at least 68°F are required for both herbs with moderate light levels for recently germinated seedlings increasing to high light levels as the plants mature. Light and temperature, as well as nutrient solution strength, are linked to the concentration of volatile compounds and essential oils in these herbs.

Applying slight stress on the plants with intense, balanced light and electrical conductivity (EC) will give improved oil content and flavor profile and is worth experimenting with particularly if winter basil quality from the store-bought product disappoints.

Electrical conductivity levels in solution culture are recommended to be maintained at 1.6-1.8 for sweet basil and cilantro and up to 2.2-2.4 for Greek and bush basil.

For those with limited space, basil and cilantro can be grown as microgreens, harvested in the seedling stage after the development of two to four leaves. Even at this delicate stage these herbs develop their characteristic flavor in a milder form well-suited to many culinary uses.

Basil or cilantro seed can be thickly sown on specifically designed microgreens mats or pads developed for hydroponic use and grown for two to three weeks on a diluted nutrient solution under moderate light levels. These are harvested by cutting at the base of the seedlings ready for use and a new crop resown.

Basil and cilantro are two of the most popular herbs grown in hydroponic indoor gardens. Giving some consideration to the wide range of cultivars, growth forms, disease resistances and different flavor profiles on offer make these exciting crops to trial as well as valuable pairings to hydroponic, vine-ripened tomatoes and spicy chillies.

Rosemary

Rosemary is a popular culinary hydroponic herblike hydroponic cilantro, hydroponic dill, hydroponic thyme, andhydroponic mint. Native to the Mediterranean, rosemary's aromatic, needle-like leaves are a highly-prized ingredient in beauty products and in culinary arts. The fresh or dried evergreen leaves are customarily used in Italian cooking as an exclusively astringent flavour-enhancer. Herbal tea is often made with the leaves of the rosemary plant. One of the best things about rosemary aquaponics and rosemary hydroponics is you can commence new crops throughout the year, providing the seed or starter plants are readily accessible. Rosemary can be raised through both seeds as well as cuttings. As previously noted, Rosemary does not propagate well using the seeds it has a very low germination rate. Thus, starting from a cutting or buying an established plant has a higher success ratio than propagating from seed.

Rosemary is a very slow-growing plant. Seed takes about 4-8 weeks to germinate. The plants take 15 months to mature, and years to grow to its full size. To speed up the process, people prefer to start new plants from the cuttings.

Rosemary Herb.

Propagating Hydroponic Rosemary

As recommended start growing rosemary from cuttings, you can get a cutting of about 10 cm long that has been cut below a leaf node on a slight angle. Rooting hormone is not compulsory, but media warmth must be maintained at 24 C you can also use a heating mat. Cutting stock may not be readily accessible year-round. But during outdoor planting season in spring and early summer you can easily fetch them from local nurseries. Make sure you strip the leaves from the lower end of the cut stem before placing it into your grow media for further growth. Wrap the lower end of the cutting in rock wool, peat, or another moisture-retaining growing media. The cuttings are geared up to transplant when roots are established and the main stem has grown past a few inches in height. This usually happens in 2–3 weeks. From transplant to the first harvest, rosemary usually takes about 6–8 weeks. While seed-raised rosemary takes about 2–3.5 weeks to germinate and an additional 3–4 weeks until they are all set to transplant.

Hydroponic Systems for Growing Rosemary

A hydroponic system is nothing but a framework which facilitates the nutrient mixture delivery to the growing plants. In case of hydroponic rosemary methods for providing the nutrient solution take account of the Passive method (which permits you to decide when and how much nutrient solution a plant needs, without installing pumps and timers), the Flood and Drain method (when trays and pots are flooded with the nutrient solution using a water pump), a Drip System (method which uses a pump to drip nutrient solution on to the plants), the Nutrient Film Technique (NFT), aeroponics, DWC, and the Kratky rosemary system are the most used systems.

Rosemary Flowering.

Growth Media for Rosemary Herb

You can grow rosemary by using hydroponics with the help of several types of growth media, including sterile root cubes made from peat and vermiculite or cellulose fibre lightweight expanded clay aggregate (L.E.C.A) and perlite. They lend a hand to support the developing delicate plants. They make it possible to grow herbs hydroponically directly in the nutrient solution, by providing some kind of physical support. Avoid using peat pots or growth cubes created for non-hydroponic gardening because they have a tendency to disintegrate and it may clog your hydroponic system causing stress to plants.

Nutrient Solution to Growing Rosemary Hydroponically

Hydroponic growers can choose from a variety of commercial nutrient solutions which is the most recommended way to go or our enthusiastic growers can also make their own solutions. One widespread recipe combines ½ ounce of potassium phosphate, 2 ounces of potassium nitrate, 3 ounces of calcium nitrate and 1½ ounce of magnesium sulfate dissolved in 25 gallons of warm water. This recipe also requires ½ pint each of iron sulfate, boric acid, and manganese chloride, as well as ½ teaspoon of zinc sulfate and copper sulfate to supply important micronutrients to the plants. Because plants exhaust the nutrient solution over time, changing the solution every three weeks is essential in order to prevent damage and stress to the plants due to unavailability of nutrient or accumulation of salts. We understand all growers are nit chemists moreover gathering above ingredients or chemicals is itself a headache to it is better to buy readymade easy to use hydroponic nutrient solution from the stores and just add it to recommended amount of water, you just need to follow the steps given on the packet rest leave it to the solution.

Growing Conditions for Hydroponic Rosemary

Growing Conditions for Rosemary.

Being a native of the warm and sunny Mediterranean area, your rosemary plants require long day length and strong uninterrupted light. For best results, you'll want rosemary in the indoor hydroponic garden to be basking under lights for a minimum of 11 hours every day. Well this can be either meet by keeping your system on widows or well-lit balcony or simply you can install high-Pressure Sodium (HPS) lamps which are best at mimicking the strong sun but it is a costly affair. You can also get away using compact fluorescent grow lamps too. If you are looking for the more budget-friendly CFL, it would be wise to keep your lights on for a couple more hours per day.

For ideal propagation, the temperature needs to be maintained at 75-84°F (24-29°C) along with 75% humidity. Once your crop has been transplanted to the hydroponic system, maintain daytime temperature and for nights at little cooler at 65°F (18°C). Its preferred pH range is 5.5-7.0, EC: 1.0–1.6, 700-1120 PPM and good air circulation is a necessity.

Harvesting Hydroponic Rosemary

Once your plant is established and is growing well, you can harvest rosemary cuttings at any time for daily culinary use. In fact, this daily (or weekly) pruning will provide you full and healthy plants as it encourages more vegetation. Keeping several plants in the system adds benefit during harvesting as you will always have one to take cuttings from. If you are cultivating Rosemary for drying purposes you could wait until the plant has just begun to flower. This is the stage when the plant has its maximum oil content and flavour. All you need is a pair of kitchen scissors and a harvesting basket. Simply cut off the top 2 to 3 inches of each stem, leaving green leaves and be careful to not cut the plant too close. You have to be sure and give it time to recover its greenery before winter sets in. You can preserve your rosemary harvest by bundling the clippings with a rubber band and hanging them upside down to air dry. This is the same method as you would use for hydroponic lavender. Once the leaves are dry, which may take about 10 days, strip them off the stems and place them in a container with a tight-fitting lid for later use.

Rose

Hydroponic flower gardens produce high-quality plants and flowers by reducing the effects of pests and disease in plant development and encourages faster growth in growing plants. Roses have been no exception to this. Many rose producers are producing roses even at the commercial level by using a hydroponic system with great success.

Types of Roses for the Hydroponic System

First, you should decide what type of roses to grow using the hydroponic technique. There are a number of different cultivars that grow well in hydroponic systems such as climbing roses (this will require installation of a trellis system), rest comes under rose bush hydroponics which comprises of shrub roses and modern roses (hybrid tea roses or long-stemmed roses). The choice of cultivar may also vary with your ultimate aim means whether you are growing for full flower or cut flower hydroponics.

Pink Roses for Hydroponics.

Propagation of Roses in Hydroponics

Planting stem cuttings is one of the most popular ways to propagate roses else rose seeds is another method of propagation but usually rose seeds are less used. Propagating from cuttings is by far the most commonly used method of reproducing roses.

They say when a flower does not blooms you fix the environment in which it grows not the flower. So growing hydroponic roses or aquaponic roses through cuttings is the best method but it is also a little bit tricky because cuttings lack a root system to take up water and essential nutrients. So, in order to propagate your roses successfully, one must create an environment that will uphold them until they build up enough roots to survive on their own.

The perfect time to take cuttings from plants is when they are growing vigorously, usually in early summer. Stems that are neither brand new nor fully mature and have fading flowers (or flowers that just have lost their petals) are advantageous. A stem cutting with a flower bud showing no color is too young. Start with filling a clean container with a best quality growing mix. It is recommended that the potting medium must be light and fast-draining, yet it should contain enough organic matter to stay moist. Quality potting soils can be purchased from your local garden stores or various sites, or you can create your own mix by combining grow media like perlite, vermiculite, and peat moss. Water the potting mixes so that it is moist but make sure it should not be soggy.

The proper methodology for propagating rose by using stem cutting:

- Choose cuttings that are a healthy and true representative of the plant.
- Take cuttings from the upper part of the plant and from the side with the help of a blade or garden knife carefully. For some reason, cuttings taken from the middle of the plant do not root well.
- Choose a stem that is four to six inches long and has at least two or three leaves attached to it.
- Use a razor blade or sharp pruning shears and make a clean slice at a 45-degree angle this will maximize the rooting area. Most cuttings root perform well if the slice is made just below a leaf node (the site where branches come out of the stem).
- Remove flowers or buds from the cutting, as well as any lower leaves attached to it. Cut the remaining leaves in half to minimize moisture loss through transpiration. Also, the presence of less foliage will maximize the amount of energy the cutting can spend on developing roots as opposed to maintaining the foliage.
- Quickly dip the bottom two inches of the cutting into a cloning solution or rooting hormone if available. Rooting hormone is not always necessary but it will greatly improve the success rate.
- Using a pencil or stick to make a small hole in the growing medium for the stem cutting to fit in and gently stuff the cutting into place.
- Roses root best in bright direct light. Set your hydroponic setup in a window and sunny balcony.

- Maintain the growing media moistened and wait until roots appear, usually in three to four weeks and transfer it to the hydroponic system.

Growing Rose Hydroponically Procedure

N.F.T. (nutrient film technique) or NFT is one of the best systems for hydroponic growing. N.F.T. is the most common and successful hydroponic growing system used in rose production.

You can easily purchase a nutrient film technique system setup from any local or online gardening store. These systems usually consist of little trays with plastic humidity covers in which you can grow your roses and feed them with water nutrient pump.

For convenience, you can also buy some rubber mats and lay them on the surface where you will place your nutrient film technique system. Because water may spill from the nutrient pump, any dampness that gets on the table or surface could cause the trays to slip off and break.

You will also have to go to a local hardware store and buy some grow lights, especially for indoor gardening, to hang grow lights over the trays. If growing your hydroponic roses in a basement or living area, expect to use the grow lights for a maximum of 12 hours every day. The areas still exposed to direct sunlight, the grow lights will just be a supplemental light source for the cloudy days. Six to eight hours of direct sunlight for growing roses is standard.

Rose Plants in Hydroponics.

Steps of Growing Rose Hydroponically

Set up your nutrient film technique system by following the directions that come along with the kit. Plug its water nutrient pump into a power strip carefully since it'll be operating at different duration every day over a period of months.

Be sure to buy healthy rose seedlings or stem cuttings that don't have any disease or insect-pest infection since hydroponic systems are hypersensitive to diseased plants and it will spread easily.

Look for net pots in any gardening store. Use some distilled water to soak up the roots of your rose seedlings or cuttings. Place the seedlings into the net pots using moistened growing medium. You can buy grow medium for hydroponic roses in most of the gardening stores.

Purchase a pH analyzer from any gardening store if one wasn't included in your hydroponic kit. These meters keep proper measurement on the acidity and alkalinity level of the growing medium and nutrients solution you're feeding to the growing plants. Roses require about 6.5 to 7.5 pH ranges in order to grow properly.

Keep a close watch on the water nutrient pump to ensure it keeps working properly especially if it is on a timer system. Refill the system with water if the level starts getting too low. Prune dead or damaged branches as the rose plants develop.

Rose bushes can become large, so you will have to give each bush at least four square feet of growing space. This permits light to reach your roses and avoid branches from becoming entangled and harming each other. Secondly, you will require some kind of support for your roses while at the same time allowing the roots to hang down in your hydroponic nutrient solution for proper nutrient intake.

Hydroponic rose needs steady moisture and air, allowing the roots to remain at a standstill in the hydroponic nutrient solution will harm the roots. You will need an aerator that pumps water continuously past the roots.

Ideal growing temperatures are between sixty-five and seventy-five degrees for hydroponic roses.

Hydroponic Nutrient for Rose

The nutrient solution to Hydroponic is just like fertilizers to the soil. Fundamentally, a Hydroponic nutrient solution is a liquid-filled with all of the essential nutrients so that plant roots can come into contact for its growth and grow well.

Hydroponic Fertilizers for Roses.

Compared to the soil, Hydroponic makes it easier to measure and fill the exact amounts of nutrients in the water solutions which fasten the nutrient uptake by developing roots. For the hydroponic rose, all you need is to log in to some authentic gardening related site and fill your cart with ready to use hydroponic rose nutrient solution and follow the steps written on its user manual. Rest the nutrients will do wonders. Just make sure you maintain the pH of the solution and do not allow nutrients to accumulate and roots to dry.

Harvesting Hydroponic Rose

Carry a clean bucket filled with fresh water when harvesting rose. This will help to keep your cut flowers fresh while harvesting full rose flowers. Use sharp bypass pruners when cutting roses. Cut rose flowers early in the morning just as the top of the bud is starting to open if you are harvesting for decoration purpose. This will maximize its vase life. Make the cut on a small angle above an outward-facing bud above a 5-leaflet leaf. The new stems that sprout from this bigger cut will be thicker and stronger. For younger plants, you may only be able to prune back up to a 3-leaflet leaf. Remove lower leaves and again cut the stem to the desired length again on a slight angle.

How to Harvest Seeds in Growing Rose Hydroponically?

Once a rose bush has bloomed rose hips are formed but not all blooms will form rose hips and several are likely deadheaded before the rose hips can truly form. Not doing any deadheading of the old rose flowers will permit the rose hips to form, which can then be harvested either to utilize the seeds inside to grow a new rose bush of your own. The rose hips are usually collected in late summer or fall once they have ripened. Some of the rose hips turn red, yellow or orange to help tell us when they have ripened. Once all of the rose hips have been harvested and collected, it is time to process the seeds present in them. Cut each rose hip open carefully using a knife and take out the seeds. Once all seeds have been removed from the rose hips, rinse the seeds off properly to remove any of the pulp from the rose hips if present on them. This is the harvesting rose seeds. You can also store your rose bush seeds in a cool, dry place for a short period of time or start right away with preparing the seeds and growing roses from the seeds.

Lavender

Lavender is one of the most admired herbs, its aromatic and herbal properties along with the appearance of its flowers has attracted many growers. A hedge of lavender in full bloom is pleasant to see and you can raise your own lavender hedge amazingly economically if you grow the lavender from the seeds.

This gardening procedure offers greater benefits compared to growing lavender in conventional soil medium. Without soil, roots obtain their nutrient solution right away, water stress is never an issue, and herbs stay small and efficient so plants are proficient to convert energy into their top growth. It's simple and, best of all; you will have a fresh, flavourful supply throughout the year.

Can lavender be grown hydroponically, yes, absolutely, we are ready to enrich your garden with wonder herb like lavender moreover growing lavender indoors hydroponically is also possible. Outdoor plants are subjected to many variables, including temperature, insect-pest attack, precipitation, and lighting. Since lavender does well in a mild, consistent and controlled environment, indoor hydroponics is a favored method of growing. Normal indoor room temperatures are ideal and can be regulated so this makes growing lavender indoors easier and less vulnerable to temperature fluctuations often found with outdoor growing environments.

Requirements to Grow Hydroponic Lavender

Requirements for Hydroponic Lavender.

- Give the plants 6 hours or more of full sun every day.
- Most people over irrigate the plants that may cause stress.
- Plants require good drainage and Aeration so don't let nutrients accumulate.
- Do not use too much of rock wool that will bring moisture to the plants.
- Prune back the plants in the late fall.
- Dampness the biggest enemy of the growing plant and at all cost dampness should be avoided to get a good harvest.

Hydroponic Lavender Seed Germination

As seed produces variable plants and takes time to propagate, so for growing lavender propagation is commonly done by cuttings of selected healthy plants. It propagates well by using softwood and semi-hardwood cuttings.

For propagating lavender plants through seed you can buy seeds of selected high yielding varieties from any gardening store. Lavender seeds germinate between 10 to 28 days after sowing. Germination of lavender seed is quite low it might go below 50 percent; for improving germination rates scarification can be done. Seed can be sown indoors six weeks before the last frost date. There are between approximately 850 and 1300 Lavender seeds present per gram, depending on the variety. You can place seeds on the grow media starter cubes and as the seed sprouts and 2-3 true leaves appear your seedling is ready to transplant. Lavender seeds require warm temperatures to germinate preferably between 65 and 70 degrees Fahrenheit (18 to 21 Celsius).

Your first year of growth will not be remarkable, but by year two, you can expect to have large, blooming lavender. That's why it said starting lavender plants from seed is not difficult, but it does have need of time, some patience.

Best Hydroponic System for Hydroponic Lavender

Herbs like lavender can be grown in a variety of hydroponic setups it depends on the scale at which herb is to be grown like for home growers Kratky lavenderis suitable because of its small size and ease of usage. Another easy system specially designed for home growers is DWC (deep water culture) which also helps in aeration using an air pump. Systems like NFT and ebb and flowsystems are usually used for outdoors and when you are looking for growing many herbs in one system.

Nutrient Solution for Hydroponic Lavender

Nutrient mix specifically designed for hydroponics are another key component of growing hydroponic lavender. You can fetch ready to use hydroponic nutrient solution for growing lavender in the hydroponic system from any gardening store online or in the market usually, bloom formulation is preferred for lavender. It is suggested to drain, clean and re-add a new nutrient solution to maintain optimal conditions for developing plants.

Other important Considerations for growing hydroponic lavender include light, temperature, humidity, and pH level. Lavender requires at the very least 6 hours of bright, unobstructed sunlight every day. A south-facing window or balcony may provide adequate lighting for growing lavender. Rotating the system to ensure all sides receive sufficient coverage is suitable. Grow lights are suggested for extra supplementation; especially if your plants are not receiving the minimum natural sunlight to remain healthy this is often encountered in indoor conditions. Standard fluorescent lamps may be sufficient, but their impact is less effective in comparison to modern LED grow lights. High-intensity discharge lights are also a beneficial especially metal halide, which is superior for growing leafy herbs like lavender.

Indoor temperatures ranging in the 70-75-degree range are ideal, along with a relative humidity level maintained around the 40-60% range. As with all hydroponic gardening, maintaining sufficient ventilation for your developing lavender is important. Oscillating fans, as well as ceiling fans also works well. The pH levels should be consistently tested to maintain a healthy range between 6.4-6.8 with 700-980 PPM.

Harvesting of Hydro Lavender

The best time to harvest fresh lavender is when the bottom flowers of each stem are just about to open. This is when the lavender is at its most vibrant and fragrant stage. Cut the flowers at the base of the stems, near to the foliage. Then, Prune the plant back to the newly grown leaves. This may encourage your plant to flower a second time, in the fall season.

If all or most of the flowers have opened on the lavender plant, then it is too late to harvest for its use as herbal purposes. If you are planning to dry the lavender for latter use, harvest it when about 3/4 of the blooms have opened.

Another tip for you if you wish to decorate your house with lavender, places the flowers in a vase, but do not put the roots in water. This just causes the flowers to fall off faster and makes the stems soggy.

Hydroponic Citrus Grove

Citrus is not a crop to be left out of the hydroponic realm. While we may be more used to thinking of expansive orchards of spreading citrus trees grown in sun-drenched climates, with a little modification citrus can be a valuable and productive specimen for the indoor garden. By selecting high quality, grafted, and dwarf varieties of citrus, a thriving crop can be grown all year indoors, avoiding the limitations of cold outdoor environments and the risk of frost damage. As an added bonus, citrus flowers are highly fragrant and the glossy, deep-green leaves make an attractive display within any hydroponic system.

Citrus Types

While a wide range of citrus is grown commercially using hydroponic methods, these are largely long-term outdoor systems with drip irrigation. Some citrus such as Makrut lime (also called k-lime), grown largely for its fragrant leaves, are produced in greenhouses where heating is applied to extend the harvest season. With a smaller indoor garden, the objective is to maximize yields and fruit quality, while at the same time growing a healthy and attractive tree. To achieve this, investing in dwarf, grafted trees of known cultivars is the best option, as although citrus can be grown from seed, the characteristics of the resulting tree are highly unpredictable. The time to first flowering is often many years and fruit is generally inferior and lower yielding than that of named varieties on grafted plants. There are even some types of grafted trees with two or three different types of citrus grafted onto the one rootstock, making more efficient use of valuable growing space. Grafting tends to produce superior trees and crops as the rootstock provides additional vigor and disease resistance. Citrus trees are great specimens for restricted spaces as many varieties, such as lemons and limes, tend to fruit relatively young and while still quite small in stature. They can also be pruned and trimmed to restrain excessive vegetative growth and kept compact while at the same time, still be highly productive. If grown with a restricted root zone volume in limited sized pots, the height and spread of the tree can be minimized and citrus can even be grown as fruiting bonsai specimens.

Hydroponic Citrus Varieties

The most commonly grown citrus for hydroponic production are the smaller types of lemons and limes; however, there are some more unusual citrus types which offer a unique opportunity to grow a tree crop indoors. These include the tiny-fruited orange kumquat, the limequat, the Australian finger lime (Citrus austalasica) blood limes, the strangely shaped Buddha's hand citrus, and the uniquely flavored calamansi. The Australian finger lime often called citrus caviar, is a relatively new specimen on the culinary scene which is currently being prized by chefs worldwide and one which is likely to become a valuable commercial hydroponic crop. Finger limes are a relative of citrus, native to Australia, but available from some citrus nurseries in the US. The fruit are elongated and finger-like, and are split open to reveal rounded lime green, tart, and juicy vesicles which resemble caviar. Finger limes require warm growing conditions and are frost sensitive outdoors, thus well suited to the year-round protected climate of an indoor garden.

For hydroponic production, suitable varieties of common citrus types include Meyer and Yen Ben lemons, Tahiti, Key and Mexican limes, mandarins, and sweet Washington navel oranges. Makrut lime is another small citrus tree ideally suited for indoor hydroponic production which produces a double crop of highly fragrant leaves which are finely sliced and used in Thai and other Asian dishes as well as small, knobbly green fruit which provide aromatic zest and juice. The advantage of Makrut lime is not only that the fresh leaves are far superior to the dried version, but the plant is often only grown for its valuable foliage and doesn't necessarily need to flower and set fruit.

Start with Plants from a Good Nursery

For a new hydroponic planting, obtaining relatively young plants from a reputable nursery is a good place to start as these can then be regularly pruned to maintain a compact and attractive shape before flower buds start to form. Obtaining older, tall and sometimes straggly plants makes the process of height control much more difficult and productivity often suffers with this type of planting stock. Once plants are in place and well established, citrus should be pruned to remove the growing points of any dominant stems and promote the production of several side branches which will allow the plant to carry more fruit and remain compact.

Light, Nutrients and Pots

Citrus, being fruiting plants, require a moderately high level of light to keep plants compact, prevent elongated growth, and produce sufficient assimilate (sugars) for high-quality fruit. A light level similar to that provided for fruiting tomato and capsicum plants is ideal, provided citrus trees are not overcrowded and kept pruned for maximum light interception. If light levels are low, citrus trees will continue to grow foliage, but often fail to flower or any sparse flowers that do open may not carry fruitlets through to maturity. Temperature optimums are also similar to many other hydroponic fruiting crops, although they vary depending on the type of citrus being grown. For optimal year-round growth, most citrus grows well within a temperature range of 75-86°F, however, will survive much cooler conditions down to 35°F for short periods of time.

Since most citrus fruits are a long-term crop and will continue to be productive for many years, growing containers of at least four gallons are recommended. Substrates such as coconut fiber or perlite are ideal, however citrus have been grown on a wide range of soilless mediums including gravel, grow rocks, peat, bark, stonewool, rice hull, and even in aeroponics. Good drainage is essential with citrus as oversaturated root zones can be prone to root rot, so the addition of some coarser material in the base of the growing container is always a good idea for long-term plants. Drip irrigation systems are most often utilized for nutrient delivery with standard vegetative and fruiting formulation applied during the different stages of citrus tree growth. A nutrient formation high in potassium with additional magnesium is essential during the fruit formation phase, as citrus fruit quality is related to potassium nutrition as well as environmental growing conditions.

Tips for Indoor Cultivation

Apart from good light and sufficient warmth, citrus benefit from some gentle air flow through the foliage. This helps remove any stale air and excess humidity and reduces the disease risk to the plants. One of the most vital aspects of growing fruiting citrus indoors is pollination which is essential for fruit set. Outdoors, bees and other insects carry out the pollen transfer role with great efficiency, however indoors this needs to be done manually. The process is relatively simple as citrus don't need to be cross pollinated. A small paint brush can be used to collect the yellow pollen grains from within a citrus flower; these are then brushed off into the stigma (found at the top of the long column in the middle of the flower). The stigma becomes sticky when receptive to pollen, allowing the brushed-on pollen grains to adhere and stay in place. This process can be carried out a few times a week when citrus trees are in flower and is usually highly successful. It is normal for many types of citrus to produce masses of highly fragrant flowers, far more than the tree could ever support as fruit. For this reason, it's not usual to see the excess flowers or tiny fruitlets drop from the tree this is just a natural self-pruning process which thins fruit to only those which the tree can support through to maturity. Young trees may only be able to carry a small handful of fruit in the first season and fruit loading will increase with the age and size of the tree.

Harvesting Hydroponic Citrus

Harvesting citrus fruit is relatively simple its best to cut these from the tree rather than pull them off as this often snaps branches or causes fruit damage. Lemons and limes, once mature, can be left on the tree for a considerable length of time, often months, before harvest and this is often used as a way to store excess fruit until it's required. However, retaining a high fruit load can delay flowering and the next crop of young fruit, so prompt removal is often a better option. Excess citrus fruit and even k-lime lime leaves can be stored in the freezer until required or processed into juice, jellies, marmalade, or syrups.

Whether it's the new and exotic Australian finger lime or a common everyday lemon tree, hydroponic citrus fruits are a valuable addition to any indoor garden. Fragrant flowers, glossy deep green foliage, and bright decorative fruit make these ideal productive specimens well suited to hydroponic systems.

Sage

Sage is an addition to the list of herbs like hydroponic rosemary, hydroponic lavenderhydroponic dill and hydroponic mint that grow successfully well with hydroponics, and can be used for medicinal or aromatherapy purposes. One of the members of mint family, culinary sage is a highly aromatic herb with a subtle, earthy characteristic flavour. Sage is a hardy perennial with attractive, grayish-green leaves that like as good in a perennial border as they do in an herbal hydroponic garden. It grows spikes of spring flowers in various colours, including purple, blue, white, and pink. Not all sage varieties are used for culinary; the most famous kitchen sage is called Salvia officinalis. The best part about raising sage is that you only need one of these incredibly easy-to-maintain plants in your garden to enjoy its flavour throughout the year. Its flavour is so strong that only a dash is needed to flavour a dish. Sage is also one of the few herbs that, even as its leaves grow up larger, the flavour intensifies. Nothing like many herbs, sage leaves are still delicious after the plant starts flowering.

Since this gifted herb is evergreen in most seasons, you can also harvest sage well into late fall. at the same time tender herbs, like basil, might die on the first freeze, sage will still be growing very well. Therefore, Growing garden sage is so economical, easy and time-saving.

How to Grow Sage?

Sage will grow almost anywhere both indoor and outdoor, but it provides the best-flavoured leaves when it receives a lot of sunlight. So you should place your hydroponic system accordingly so that your plants receive adequate sunlight.

Sage can prove little challenging when raised by using seeds, but it is extremely easy to grow from cuttings or by "layering". You can purchase your first sage plants from any garden center, and later you can propagate new plants through one of the two methods listed below. in spite of which propagation method you choose, plant young sage plants only after the temperature hits 65°F, one to two weeks prior to the last frost.

Growing Sage without Soil.

Propagate from cuttings: Clip a three-inch cutting from the very tip of a sage plant stem, apply rooting hormone if available to the exposed portion of the stem cutting, and plant it in either coco-peat or vermiculite grow media starter. Roots will come into sight within six weeks. After that, you can transfer to the system. Perlite is an ideal growth media for sage because of the drainage quality it offers for hydroponics.

Seed Germination in Growing Sage Hydroponically

Propagation through seed is a less preferred method for raising hydroponic sage plants. Sage seeds are slow growers and may take between 10 to 21 days to germinate. Its Seed has poor storage and germination. When raised from seeds, sage takes about 2 years to reach its mature size. That is why most of the home gardeners prefer and recommend starting culinary sage from cuttings or divisions utilizing the outer or newer growth. If starting seeds indoors, you should sow it under grow lights about 6-8 weeks prior to the last expected frost.

Hydroponic System for Growing Sage

Even though there are a lot of different types of hydroponic systems, they all work using the same basic principle. First and foremost, hydroponic growing systems permit plant roots to come into direct contact with nutrients and oxygen, which is both necessary to plant growth, devoid of using soil. Instead of soil, some hydroponic systems use different types of growing media, like stone wool, clay pebbles, coco coir, vermiculite or perlite to extend support for a plant's developing roots. Bottom line is for growing hydroponic sage you can use a variety of hydroponic systems like DWC, Kratky, ebb, and flow and for having a whole sole hydroponic herb garden you can use a system like NFT. For commercial cultivation system like aeroponics, that does not involve any growing media can be used.

Nutrient Solution for Hydroponic Sage

As for plant nutrients, these are delivered directly to the plant roots through a water-based, nutrient-rich solution having adequate nutrients. This solution can be applied to the roots using several different methods like via a water pump. Some of these methods involve but are not limited to, the developing roots being suspended in the nutrient solution, such as the deepwater culture method, or the roots can also be misted with the specialized nutrient solution, which is utilized in aeroponics. You can these specially formulated nutrients from various online hydroponic platforms along with the user manual with clear cut instructions. Stock up the nutrient solution every two weeks or so when it becomes a little cloudy or accumulation of salts begins else they will cause nutrient stress and deoxygenated conditions.

Growing Conditions for Hydroponic Sage

Light, along with nutrients and oxygen, must be accessible for plants to absorb, as it is a vital component required for photosynthesis. A lot like the nutrient solution, appropriate amounts of light can be delivered to sage plants raised hydroponically in a few different ways. Some of the hydroponic systems are open, relying on natural lighting, whereas other systems depend on artificial lighting (grow lights) to offer plants with sufficient amounts of light. The sage plant must have about 12 hours of light every day, so use grow lights if necessary like in indoor conditions. One must regularly test the nutrient system's pH with the help of testing strips. Hydroponic sage must have a consistent pH of around 5.5 to 6.0. Preferred temperature for the day is 75-85°Farenheit and for the night it is 60-80° Farhenheit.

Harvesting Hydroponic Sage

During the first year, harvest less to ensure that the plant grows fully to its vigour. After the first year of growth, be sure to leave a few stems so that the sage plant can rejuvenate. Once your plant is fully established, one plant can be harvested up to three times in one growth season. Avoid practicing harvesting in the fall so the plant can prepare itself for the winter. It is recommended to substitute the plants every few years so that they remain productive. Sage can be harvested whenever needed, by clipping just above the place where two leaves meet. For the highest concentration of their aromatic oils, harvest sage in the morning, once the dew has dried.

Harvested Sage Herb.

It is also suggested to conduct a larger sage harvest about twice during its growing season, in order to support a prolific, properly shaped and maintained plant. You just have to cut the sage stems back, harvesting not more than half of the plant.

References

- WhatCanYouGrowHydroponically: modularhydro.com, Retrieved 13 July, 2020
- Aromatic-adventures-growing-exotic-herbs-and-spices-hydroponically-1142: maximumyield.com, Retrieved 14 March, 2020
- Fast-growing-hydroponic: growgardener.com, Retrieved 29 April, 2020
- Growing-mint-hydroponically-pudina-planting-care: gardeningtips.in, Retrieved 31 March, 2020
- Hydroponic-ginger-turmeric-production-2840: maximumyield.com, Retrieved 18 June, 2020
- Growing-oregano-in-hydroponics: university.upstartfarmers.com, Retrieved 15 July, 2020
- Growing-rose-hydroponically-hydroponic-rose-nutrients: gardeningtips.in, Retrieved 05 April, 2020
- Cultivating-a-hydroponic-citrus-grove-17503: maximumyield.com, Retrieved 16 August, 2020

5

Equipment used in Hydroponic Farming

Different types of equipment are used in hydroponic farming for improving the quality of the hydroponic plants. Some of them are leaf sensor, spray nozzle, water chiller, net-pot, mesh pot, etc. The topics elaborated in this chapter will help in gaining a better perspective of the equipment used in hydroponic farming.

Nutrient Doser

A nutrient doser is a system of equipment that helps maintain pH and nutrient levels in a hydroponics system. Some nutrient dosers come equipped to monitor four or more solutions and with probes that measure the reservoir of the fertilizer or other chemicals.

This tool initiates automatic dosing at pre-configured levels. An automatic nutrient doser ensures stable chemical levels, prevents overcorrecting, saves on labor, and avoids shocking the plants from a sudden change to their environment. Nutrient dosers promote healthy and steady plant growth.

Nutrient dosers work by pulling nutrients into an injector and mixing them with water to produce precise parts per million (ppm) or electrical conductivity (EC) levels. A nutrient doser is usually set up in a garden that has several wall-mounted injectors, with each of them having a tube that runs into a container of raw nutrient. As several parts of a nutrient regimen cannot be premixed due to adverse reactions, nutrient dosers keep all the chemical parts separate until they are fed into the water line.

Automatic nutrient dosers help to avoid human errors and in most cases can create higher yields than plants grown through a manual dosing system.

Automated Dosing System

An automated dosing system really does two things: Monitoring and dosing.

Monitoring

Using a probe positioned in your nutrient reservoir, the system measures your pH, EC levels, and water temperature.

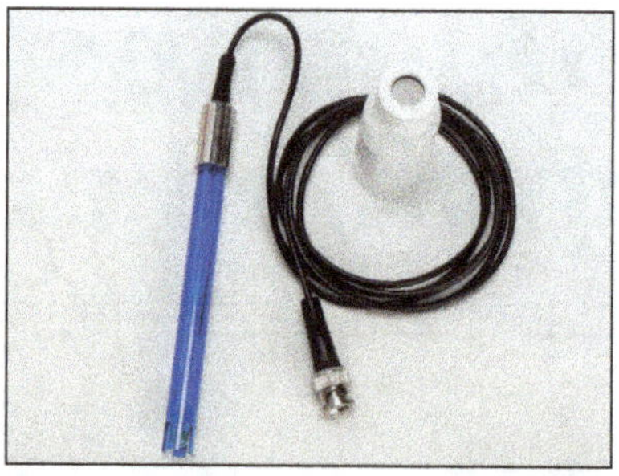

These are the most critical variables for your hydroponic growing system. (If you're a beginner, EC stands for electrical conductivity, which uses conductivity to measure the salts in the system. The level of salts present correlates to the level of dissolved nutrients in the water).

Dosing

The dosing system powers several pumps to add nutrients or minerals to your system.

Dosing systems can save significant time and labor, as well as protecting against crop shock events; both goals are appropriate for both large and small systems. In fact, small systems can benefit from automated dosing just as much as large systems, since their nutrient levels usually fluctuate more widely than large systems.

What are the Benefits of Auto Dosing?

The most obvious benefit of automatic dosing is that you won't have to spend time testing and adjusting your pH, temperature and EC levels, but the implications of that benefit reach beyond that.

Save on Labor

Many growers spend 15-30 minutes a day testing and correcting their system levels. (2-3 hours per week) This means testing for pH levels, EC levels, and other nutrients as needed. Beginning growers will often spend more time on this until they get a feel for nutrient levels.

Avoid over Correcting

Many folks unintentionally over-correct one or two of their variables. This is kind of like adding too much flour to the pancake mix. You have to add more water to balance out the flour, but then if you add too much water you have to add more flour. Auto-dosing systems help keep these levels stable, and provides the optimal environment for your plants. This results in bigger, healthier plants.

Avoid Shocking your Plants

Plant shock events can occur whenever a plant experiences a sudden change in environment or nutrient flow.

One common scenario happens when the water is dumped from the system. Some growers do this weekly to start fresh at the beginning of the week with correct nutrient levels. This can be harmful to your plants (as well as being wasteful).

The reason for this is that the water in the bottom of the tank is usually highly concentrated, with more nutrients left from evaporation throughout the week. After the tank is emptied, fresh water is added with nutrients at the correct levels. When the fresh water (with correct nutrients) is added back to the system, the plants absorb it.

The difference is that now this is a more diluted solution than before. Since plants take up water primarily through osmosis, which uses a diffusion gradient, a change in concentration changes that gradient. Before adjusting, the plants take in a larger-than-normal amount of the diluted solution. This is called "flooding" your plants. It is usually a minor stress on the plants, although if nutrient levels fluctuate too wildly you can shock your plant and even cause crop loss.

The way an auto dosing system corrects for this is by measuring and adding nutrients in real time. The system can take measurements on a scheduled interval. It will decrease nutrient levels by adding water, or, conversely, increase nutrient levels by adding nutrients to the reservoir.

This level of control over your system means that you will dump your system less frequently, which saves you the cost of lost nutrients and results in less waste. The consistent nutrient supply will result in healthier, more robust plants.

Keep your Freedom to Vacate

When a farmer has to be present all day, every day, to take care of his (or her) pH and nutrient levels, it tends to be a bit binding. Having an auto dosing system can give farmers the freedom to leave the farm for a day or two. It makes leaving the farm in the care of another person easier in case they want to leave for longer as well.

For Aquaponic Growers: Maintain Iron and pH Levels

Aquaponic growers don't use nutrient solutions. However, many AP growers will still consider an auto dose system for checking for and correcting pH and iron.

How does it Work?

When you set up the automated system, you will calibrate your unit to your desired pH, EC and temperature. You will then connect your inputs to the system. For example, we have a reservoir for each of supplements; part A (main nutrient mix), part B (calcium nitrate with some magnesium sulfate (Epsom salt)), and pH down.

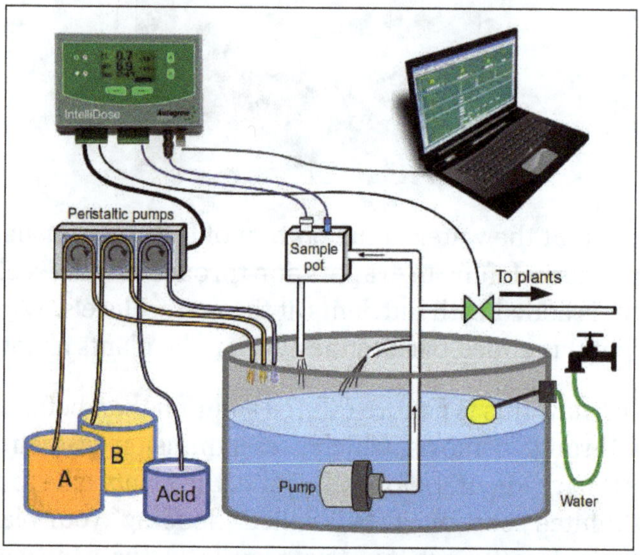

These will usually be kept in five-gallon buckets with a small hose to connect to the dosing system. The automated system will take a test every so often, and then, using one of the small pumps that goes with the system, it pulls the supplement from the bucket into your system.

Grow Light

Grow lights either attempt to provide a light spectrum similar to that of the sun, or to provide a spectrum that is more tailored to the needs of the plants being cultivated. Outdoor conditions are mimicked with varying colour, temperatures and spectral outputs from the grow light, as well as varying the intensity of the lamps. Depending on the type of plant being cultivated, the stage of cultivation (e.g. the germination/vegetative phase or the flowering/fruiting phase), and the photoperiod required by the plants, specific ranges of spectrum, luminous efficacy and colour temperature are desirable for use with specific plants and time periods.

Russian botanist Andrei Famintsyn was the first to use artificial light for plant growing and research.

Typical Use

Grow lights are used for horticulture, indoor gardening, plant propagation and food production, including indoor hydroponics and aquatic plants. Although most grow lights are used on an industrial level, they can also be used in households.

According to the inverse-square law, the intensity of light radiating from a point source (in this case a bulb) that reaches a surface is inversely proportional to the square of the surface's distance from the source (if an object is twice as far away, it receives only a quarter the light) which is a serious hurdle for indoor growers, and many techniques are employed to use light as efficiently as possible. Reflectors are thus often used in the lights to maximize light efficiency. Plants or lights are moved as close together as possible so that they receive equal lighting and that all light coming from the lights falls on the plants rather than on the surrounding area.

Example of an HPS grow light set up in a grow tent. The setup includes a carbon filter to remove odors, and ducting to exhaust hot air using a powerful exhaust fan.

A range of bulb types can be used as grow lights, such as incandescents, fluorescent lights, high-intensity discharge lamps (HID), and light-emitting diodes (LED). Today, the most widely used lights for professional use are HIDs and fluorescents. Indoor flower and vegetable growers typically use high-pressure sodium (HPS/SON) and metal halide (MH) HID lights, but fluorescents and LEDs are replacing metal halides due to their efficiency and economy.

Metal halide lights are regularly used for the vegetative phase of plant growth, as they emit larger amounts of blue and ultraviolet radiation. With the introduction of ceramic metal halide lighting and full-spectrum metal halide lighting, they are increasingly being utilized as an exclusive source of light for both vegetative and reproductive growth stages. Blue spectrum light may trigger a greater vegetative response in plants.

High-pressure sodium lights are also used as a single source of light throughout the vegetative and reproductive stages. As well, they may be used as an amendment to full-spectrum lighting during the reproductive stage. Red spectrum light may trigger a greater flowering response in plants. If high-pressure sodium lights are used for the vegetative phase, plants grow slightly more quickly, but will have longer internodes, and may be longer overall.

In recent years LED technology has been introduced into the grow light market. By designing an indoor grow light using diodes, specific wavelengths of light can be produced. NASA has tested LED grow lights for their high efficiency in growing food in space for extraterrestrial colonization. Findings showed that plants are affected by light in the red, green and blue parts of the visible light spectrum.

Common Types

High Intensity Discharge Lights

Fluorescent lighting was formerly the most common type of indoor grow light but HID lights have surpassed them. High intensity discharge lamps have a high lumen-per-watt efficiency. There are several different types of HID lights including mercury vapor, metal halide, high pressure sodium and conversion bulbs. Metal halide and HPS lamps produce a color spectrum that is somewhat comparable to the sun and can be used to grow plants. Mercury vapor lamps were the first type of HIDs and were widely used for street lighting, but when it comes to indoor gardening they produce a relatively poor spectrum for plant growth so they have been mostly replaced by other types of HIDs for growing plants.

All HID grow lights require an electrical ballast to operate, and each ballast has a particular power rating. Popular HID ratings include 150W, 250W, 400W, 600W and 1000W. 600W HID lights are the most electrically efficient as far as light produced, followed by 1000W. A 600W HPS produces 7% more light (lumen-per-watt) than a 1000W HPS.

Although all HID lamps work on the same principle, the different types of bulbs have different starting and voltage requirements, as well as different operating characteristics and physical shape. Because of this a bulb won't work properly without a matching ballast, even if the bulb will physically screw in. In addition to producing lower levels of light, mismatched bulbs and ballasts will stop working early, or may even burn out immediately.

Metal Halide

Metal halide bulbs are a type of HID light that emit light in the blue and violet parts of the light spectrum, which is similar to the light that is available outdoors during spring. Because their light mimics the color spectrum of the sun, some growers find that plants look more pleasing under a metal halide than other types of HID lights such as the HPS which distort the color of plants. Therefore, it's more common for a metal halide to be used when the plants are on display in the home (for example with ornamental plants) and natural color is preferred. Metal halide bulbs need to be replaced about once a year, compared to HPS lights which last twice as long.

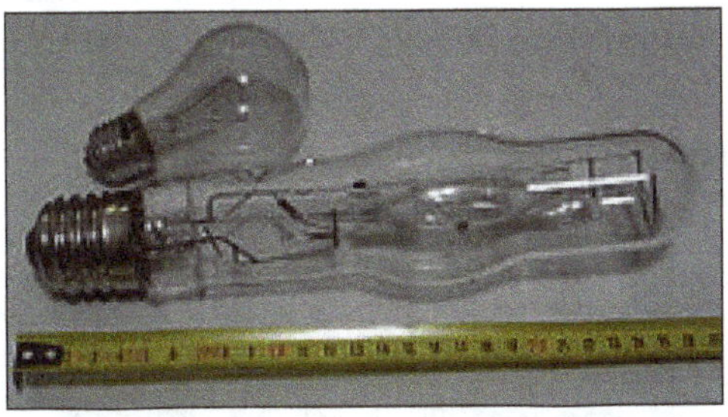

400W Metal halide bulb compared to smaller incandescent bulb.

Metal halide lamps are widely used in the horticultural industry and are well-suited to supporting plants in earlier developmental stages by promoting stronger roots, better resistance against disease and more compact growth. The blue spectrum of light encourages compact, leafy growth and may be better suited to growing vegetative plants with lots of foliage.

A metal halide bulb produces 60-125 lumens/watt, depending on the wattage of the bulb.

They are now being made for digital ballasts in a pulse start version, which have higher electrical efficiency (up to 110 lumens per watt) and faster warmup. One common example of a pulse start metal halide is the ceramic metal halide (CMH). Pulse start metal halide bulbs can come in any desired spectrum from cool white (7000 K) to warm white (3000 K) and even ultraviolet-heavy (10,000 K).

Ceramic Metal Halide

Ceramic metal halide (CMH) lamps are a relatively new type of HID lighting, and the technology is referred to by a few names when it comes to grow lights, including ceramic discharge metal halide (CDM), ceramic arc metal halide.

Ceramic metal halide lights are started with a pulse-starter, just like other "pulse-start" metal halides. The discharge of a ceramic metal halide bulb is contained in a type of ceramic material known as polycrystalline alumina (PCA), which is similar to the material used for an HPS. PCA reduces sodium loss, which in turn reduces color shift and variation compared to standard MH bulbs. Horticultural CDM offerings from companies such as Philips have proven to be effective sources of growth light for medium-wattage applications.

Combination MH and HPS

Combination HPS/MH lights combine a metal halide and a high-pressure sodium in the same bulb, providing both red and blue spectrums in a single HID lamp. The combination of blue metal halide light and red high-pressure sodium light is an attempt to provide a very wide spectrum within a single lamp. This allows for a single bulb solution throughout the entire life cycle of the plant, from vegetative growth through flowering. There are potential tradeoffs for the convenience of a single bulb in terms of yield. There are however some qualitative benefits that come for the wider light spectrum.

High-pressure Sodium

An HPS (High Pressure Sodium) grow light bulb in an air-cooled reflector with hammer finish. The yellowish light is the signature color produced by an HPS.

High-pressure sodium lights are a more efficient type of HID lighting than metal halides. HPS bulbs emit light in the yellow/red visible light as well as small portions of all other visible light. Since HPS grow lights deliver more energy in the red part of the light spectrum, they may promote blooming and fruiting. They are used as a supplement to natural daylight in greenhouse lighting and full-spectrum lighting(metal halide) or, as a standalone source of light for indoors/grow chambers.

HPS grow lights are sold in the following sizes: 150W, 250W, 400W, 600W and 1000W. Of all the sizes, 600W HID lights are the most electrically efficient as far as light produced, followed by 1000W. A 600W HPS produces 7% more light (watt-for-watt) than a 1000W HPS.

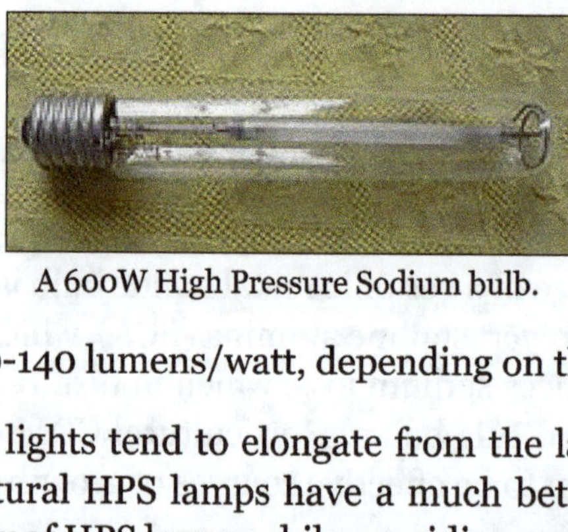

A 600W High Pressure Sodium bulb.

An HPS bulb produces 60-140 lumens/watt, depending on the wattage of the bulb.

Plants grown under HPS lights tend to elongate from the lack of blue/ultraviolet radiation. Modern horticultural HPS lamps have a much better adjusted spectrum for plant growth. The majority of HPS lamps while providing good growth, offer poor color rendering index (CRI) rendering. As a result, the yellowish light of an HPS can make monitoring plant health indoors more difficult. CRI isn't an issue when HPS lamps are used as supplemental lighting in greenhouses which make use of natural daylight (which offsets the yellow light of the HPS).

HPS grow lights suspended over the grass pitch at Tottenham Hotspur Stadium to encourage grass growth.

High-pressure sodium lights have a long usable bulb life, and six times more light output per watt of energy consumed than a standard incandescent grow light. Due to their high efficiency and the fact that plants grown in greenhouses get all the blue light they need naturally, these lights are the preferred supplemental greenhouse lights. But, in the higher latitudes, there are periods of the year where sunlight is scarce, and additional sources of light are indicated for proper growth. HPS lights may cause distinctive infrared and optical signatures, which can attract insects or other species of pests; these may in turn threaten the plants being grown. High-pressure sodium lights emit a lot of heat, which can cause leggier growth, although this can be controlled by using special air-cooled bulb reflectors or enclosures.

Conversion Bulbs

Conversion bulbs are manufactured so they work with either a MH or HPS ballast. A grower can run an HPS conversion bulb on a MH ballast, or a MH conversion bulb on a HPS ballast. The difference between the ballasts is an HPS ballast has an igniter which ignites the sodium in an HPS bulb, while a MH ballast does not. Because of this, all electrical ballasts can fire MH bulbs, but only a Switchable or HPS ballast can fire an HPS bulb without a conversion bulb. Usually a metal halide conversion bulb will be used in an HPS ballast since the MH conversion bulbs are more common.

Switchable Ballasts

A switchable ballast is an HID ballast can be used with either a metal halide or an HPS bulb of equivalent wattage. So a 600W Switchable ballast would work with either a 600W MH or HPS. Growers use these fixtures for propagating and vegetatively growing plants under the metal halide, then switching to a high-pressure sodium bulb for the fruiting or flowering stage of plant growth. To change between the lights, only the bulb needs changing and a switch needs to be set to the appropriate setting.

Light Emitting Diodes

LED grow lights are composed of multiple individual light-emitting diodes, usually in a casing with a heat sink and built-in fans. LED grow lights do not usually require a separate ballast and can be plugged directly into a standard electrical socket.

Two plants growing under an LED grow light.

Individual LEDs usually provide only a single narrow range of colors, and so different color LEDs are mixed in grow lights in proportions depending on the intended use. It is known from the study of photomorphogenesis that green, red, far-red and blue light spectra have an effect on root formation, plant growth, and flowering, but there are not enough scientific studies or field-tested trials using LED grow lights to recommend specific color ratios for optimal plant growth under LED grow lights. It has been shown that many plants will grow normally if given both red and blue light. However, many studies indicate that red and blue light only provides the most cost efficient method of growth, plant growth is still better under light supplemented with green.

White LED grow lights provide a full spectrum of light designed to mimic natural light, providing plants a balanced spectrum of red, blue and green. The spectrum used varies, however, white LED grow lights are designed to emit similar amounts of red and blue light with the added green light to appear white. White LED grow lights are often used for supplemental lighting in home and office spaces.

A large number of plant species have been assessed in greenhouse trials to make sure plants have higher quality in biomass and biochemical ingredients even higher or comparable with field conditions. Plant performance of mint, basil, lentil, lettuce, cabbage, parsley, carrot were measured by assessing health and vigor of plants and success in promoting growth. Promoting in profuse flowering of select ornamentals including primula, marigold, stock were also noticed.

In tests conducted by Philips Lighting on LED grow lights to find an optimal light recipe for growing various vegetables in greenhouses, they found that the following aspects of light affects both plant growth (photosynthesis) and plant development (morphology): light intensity, total light over time, light at which moment of the day, light/dark period per day, light quality (spectrum), light direction and light distribution over the plants. However it's noted that in tests between tomatoes, mini cucumbers and bell peppers, the optimal light recipe was not the same for all plants, and varied depending on both the crop and the region, so currently they must optimize LED lighting in greenhouses based on trial and error. They've shown that LED light affects disease resistance, taste and nutritional levels, but as of 2014 they haven't found a practical way to use that information.

Ficus plant grown under a white LED grow light.

The diodes used in initial LED grow light designs were usually 1/3 watt to 1 watt in power. However, higher wattage diodes such as 3 watt and 5 watt diodes are now commonly used in LED grow lights. For highly compacted areas, COB chips between 10 watts and 100 watts can be used. Because of heat dissipation, these chips are often less efficient.

To prevent leaf burn, LED grow lights should be kept between 12 inches (30 cm) away from plants for lower wattage lamps (under 300 watts) up to 36 inches (91 cm) away from plants for higher wattage lamps (1000 watts or more).

Historically, LED lighting was very expensive, but costs have greatly reduced over time, and their longevity has made them more popular. LED grow lights are often priced higher, watt-for-watt, than other LED lighting, due to design features that help them to be more energy efficient and last longer. In particular, because LED grow lights are relatively high power, LED grow lights are often equipped with cooling systems, as low temperature improves both the brightness and longevity. LEDs usually last for 50,000 - 90,000 hours until LM-70 is reached.

Fluorescent

Fluorescent lights come in many form factors, including long, thin bulbs as well as smaller spiral shaped bulbs (compact fluorescent lights). Fluorescent lights are available in color temperatures ranging from 2700 K to 10,000 K. The luminous efficacy ranges from 30 lm/W to 90 lm/W. The two main types of fluorescent lights used for growing plants are the tube-style lights and compact fluorescent lights.

Fluorescent grow light.

Tube-style Fluorescent Lights

Fluorescent grow lights are not as intense as HID lights and are usually used for growing vegetables and herbs indoors, or for starting seedlings to get a jump start on spring plantings. A ballast is needed to run these types of fluorescent lights.

Standard fluorescent lighting comes in multiple form factors, including the T5, T8 and T12. The brightest version is the T5. The T8 and T12 are less powerful and are more suited to plants with lower light needs. High-output fluorescent lights produce twice as much light as standard fluorescent lights. A high-output fluorescent fixture has a very thin profile, making it useful in vertically limited areas.

Fluorescents have an average usable life span of up to 20,000 hours. A fluorescent grow light produces 33-100 lumens/watt, depending on the form factor and wattage.

Compact Fluorescent Lights

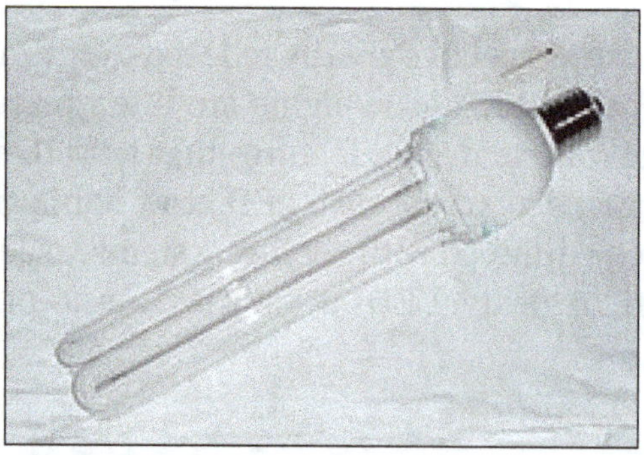

Dual spectrum compact fluorescent grow light. Actual length is about 40 cm (16 in).

Standard Compact Fluorescent Light.

Compact Fluorescent lights (CFLs) are smaller versions of fluorescent lights that were originally designed as pre-heat lamps, but are now available in rapid-start form. CFLs have largely replaced incandescent light bulbs in households because they last longer and are much more electrically efficient. In some cases, CFLs are also used as grow lights. Like standard fluorescent lights, they are useful for propagation and situations where relatively low light levels are needed.

While standard CFLs in small sizes can be used to grow plants, there are also now CFL lamps made specifically for growing plants. Often these larger compact fluorescent bulbs are sold with specially designed reflectors that direct light to plants, much like HID lights. Common CFL grow lamp sizes include 125W, 200W, 250W and 300W.

Unlike HID lights, CFLs fit in a standard mogul light socket and don't need a separate ballast.

Compact fluorescent bulbs are available in warm/red (2700 K), full spectrum or daylight (5000 K) and cool/blue (6500 K) versions. Warm red spectrum is recommended for flowering, and cool blue spectrum is recommended for vegetative growth.

Usable life span for compact fluorescent grow lights is about 10,000 hours. A CFL produces 44-80 lumens/watt, depending on the wattage of the bulb.

Examples of lumens and lumens/watt for different size CFLs:

CFL Wattage	Initial Lumens	Lumens/watt
23W	1,600	70
42W	2,800	67
85W	4,250	50
125W	7,000	56
200W	10,000	50

Cold Cathode Fluorescent Light

A cold cathode is a cathode that is not electrically heated by a filament. A cathode may be considered "cold" if it emits more electrons than can be supplied by thermionic emission alone. It is used in gas-discharge lamps, such as neon lamps, discharge tubes, and some types of vacuum tube. The other type of cathode is a hot cathode, which is heated by electric current passing through a filament. A cold cathode does not necessarily operate at a low temperature: it is often heated to its operating temperature by other methods, such as the current passing from the cathode into the gas.

Light Requirements of Plants

The quantity, quality, and duration of light regulate plant growth and development. In general, if a plant does not get enough light, it will become stunted, have reduced pigmentation, or begin shade-avoidance response. A plant that does not receive the right quality of light may exhibit physiological differences when compared to the same plants grown under optimal lighting conditions.

Grow light quantity and quality has been technologically limited in the past. High pressure sodium (HPS) and metal halide (MH) were and are still common supplemental lighting options for greenhouses and some sole-source operations. Older LED grow lights composed solely of blue and red LEDs due to both their efficiency at converting electricity to photons and efficiency at driving photosynthesis. As LEDs become less expensive and more efficient, an interest in studying light quality has risen in the field of plant science.

Light Quantity

Light quantity refers to the amount of light a plant requires each day for optimal growth. Historically, light quantity was expressed in units of W m^{-2}, lumens, or lux. While these units are useful in energy calculations, W m^{-2}, or in human lighting (lumens and lux), plant scientists now prefer to measure the photosynthetic photon flux density (PPFD), in units of $\mu mol\ m^{-2}s^{-1}$. PPFD is an explicit measure of the quantity of photons hitting a surface per square meter per second, a more accurate way to measure how plants interact with photons.

Another useful way to measure light quantity is through the daily light integral, or DLI. The DLI takes into account the PPFD and the total number of hours a plant is exposed to that PPFD to get the total quantity of photons per day, in units of mol $m^{-2}d^{-1}$. The equation for converting PPFD to DLI, assuming constant PPFD, is below.

DLI (mol $m^{-2}d^{-1}$) =0.0036 * PPFD ($\mu mol\ m^{-2}s^{-1}$) *Hours of Light.

Hortipower spectrum meter to measure grow light for plants.

The light quantity requirements for crops vary, in general the light requirement for a specific crop is greater for crops that are fruiting and flowering and is less for crops that remain vegetative. Leafy greens such as lettuce, spinach, and kale are typically considered low-light crops, requiring a DLI between 12 and 17 mol $m^{-2}d^{-1}$. Tomatoes, cucumbers, and peppers require between 20-30 mol $m^{-2}d^{-1}$. Cannabis has one of the highest light requirements of cultivated plants, requiring a DLI of up to 40 mol $m^{-2}d^{-1}$.

Light Quality

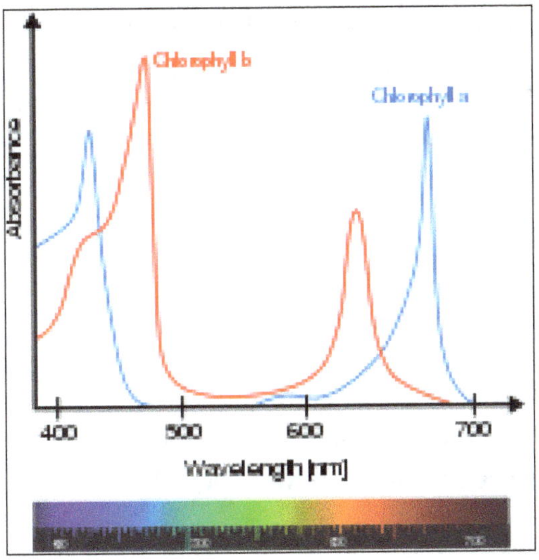

Absorbance spectra of free chlorophyll a (blue) and b (red) in a solvent. The action spectra of chlorophyll molecules are slightly modified in vivo depending on specific pigment-protein interactions.

Light quality refers to the spectral distribution of light given to a plant. Light quality is grouped into colors based on wavelength; 320-400 nanometers (nm) is UVA, 400-500 nm is blue, 500-600 nm is green, 600-700 nm is red, and 700-750 nm is far red, sometimes referred to as near-infrared. Light quality can also be expressed as ratios, e.g. 3:2 red:blue ratio, or sometimes as their peak irradiance, e.g. 450 nm blue light and 660 nm red light. Photomorphogenesis is the term for light-mediated plant responses to light spectrum. Plants are able to sense parts of the electromagnetic spectrum through a network of photoreceptors including phytochromes, cryptochromes, phototropin, and zeiltupe. Each receptor is able to sense different parts of the electromagnetic spectrum. Information about the light spectrum can affect seed germination, the signal to transition from vegetative to flowering, and the production of secondary metabolites such as anthocyanins.

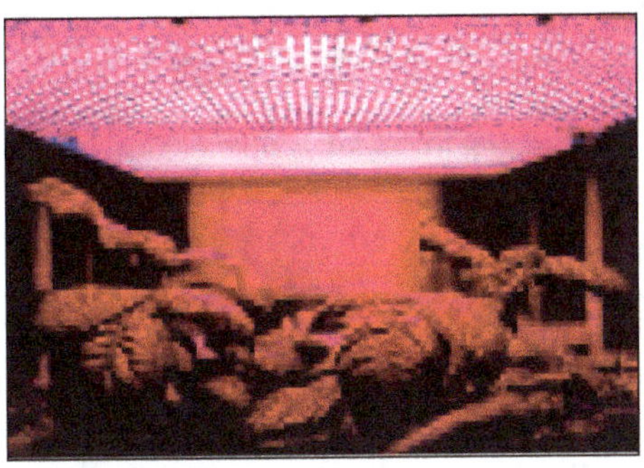

LED panel light source used in an experiment on potato plant growth by NASA.

Photoperiodism

In addition, many plants also require both dark and light periods, an effect known as photoperiodism, to trigger flowering. Therefore, lights may be turned on or off at set times. The optimum photo/dark period ratio depends on the species and variety of plant, as some prefer long days and short nights and others prefer the opposite or intermediate "day lengths".

Much emphasis is placed on photoperiod when discussing plant development. However, it is the number of hours of darkness that affects a plant's response to day length. In general, a "short-day" is one in which the photoperiod is no more than 12 hours. A "long-day" is one in which the photoperiod is no less than 14 hours. Short-day plants are those that flower when the day length is less than a critical duration. Long-day plants are those that only flower when the photoperiod is greater than a critical duration. Day-neutral plants are those that flower regardless of photoperiod.

Plants that flower in response to photoperiod may have a facultative or obligate response. A facultative response means that a plant will eventually flower regardless of photoperiod, but will flower faster if grown under a particular photoperiod. An obligate response means that the plant will only flower if grown under a certain photoperiod.

Photosynthetically Active Radiation

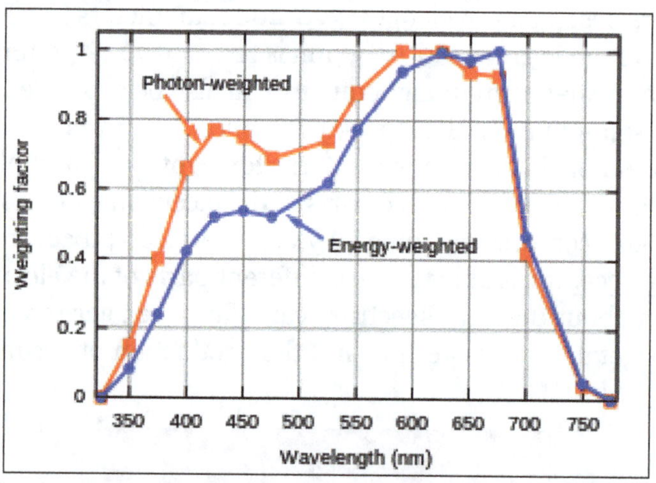

Weighting factor for photosynthesis. The photon-weighted curve is for converting PPFD to YPF; the energy-weighted curve is for weighting PAR expressed in watts or joules.

Lux and lumens are commonly used to measure light levels, but they are photometric units which measure the intensity of light as perceived by the human eye.

The spectral levels of light that can be used by plants for photosynthesis is similar to, but not the same as what's measured by lumens. Therefore, when it comes to measuring the amount of light available to plants for photosynthesis, biologists often measure the amount of photosynthetically active radiation (PAR) received by a plant. PAR designates the spectral range of solar radiation from 400 to 700 nanometers, which generally corresponds to the spectral range that photosynthetic organisms are able to use in the process of photosynthesis.

The irradiance of PAR can be expressed in units of energy flux (W/m^2), which is relevant in energy-balance considerations for photosynthetic organisms. However, photosynthesis is a quantum process and the chemical reactions of photosynthesis are more dependent on the number of photons than the amount of energy contained in the photons. Therefore, plant biologists often quantify PAR using the number of photons in the 400-700 nm range received by a surface for a specified amount of time, or the Photosynthetic Photon Flux Density (PPFD). This is normally measured using mol $m^{-2}s^{-1}$, but the value relevant for plant growth is the Daily light integral (DLI), the PPFD integrated over 24 hours. Most plant species will grow well with a DLI of 5-15 mol m^{-2} day^{-1}. Shade-tolerant species can grow with DLI values of 1-3 mol m^{-2} day^{-1}, light-demanding species easily handle 30-50 mol m^{-2} day^{-1}.

Leaf Sensor

A leaf sensor is a phytometric device (measurement of plant physiological processes) that measures water loss or the water deficit stress (WDS) in plants by real-time monitoring the moisture level in plant leaves. The first leaf sensor was developed by LeafSens, an Israeli company granted a US patent for a mechanical leaf thickness sensing device in 2001. LeafSen has made strides incorporating their leaf sensory technology into citrus orchards in Israel. A solid state smart leaf sensor technology was developed by the University of Colorado at Boulder for NASA in 2007. It was designed to help monitor and control agricultural water demand. AgriHouse received a National Science Foundation (NSF) STTR grant in conjunction with the University of Colorado to further develop the solid state leaf sensor technology for precision irrigation control in 2007.

Precision Monitoring

Water Deficit Stress Measurements

A Phase I research grant from the National Science Foundation in 2007 showed that the leaf sensor technology has the potential to save between 30% and 50% of irrigation water by reducing irrigation from once every 24 hours to about every 2 to 2.5 days by sensing impending water deficit stress. Leaf sensor technology developed by AgriHouse indicates water deficit stress by measuring the turgor pressure of a leaf, which decreases dramatically at the onset of leaf dehydration. Early detection of impending water deficit stress in plants can be used as an input parameter for precision irrigation control by allowing plants to communicate water requirements directly to humans and electronic interfaces. For example, a base system utilizing the wirelessly transmitted information of several sensors appropriately distributed over various sectors of a round field irrigated by a center-pivot irrigation system could tell the irrigation lever exactly when and what field sector needs to be irrigated.

AgriHouse Smart Leaf Sensor (SG-1000).

Irrigation Control

In a 2008 USDA sponsored field study AgriHouse's SG-1000 Leaf Sensor attached to dry beans demonstrated a 25% savings of irrigation water and pumping cost. In 2010 the University of Colorado, Boulder, Colorado granted AgriHouse Inc an exclusive license of its patented leaf sensor technology.

Precision irrigation monitoring using the SG-1000 leaf sensor and commercial data loggers for irrigation control has been achieved in recent years. Researchers have found a direct correlation between leaf thickness and Relative Water Content (RWC) of plant leaves using the SG-1000 Leaf Sensor under field conditions.

Water and Energy Conservation

The agriculture sustainability benefits of water and energy savings have been established using the SG-1000 leaf sensor under field conditions and in greenhouses. Plant science researchers and agronomists have utilized the SG-1000 Leaf Sensor for studying the relationship between water content and leaf cell turgidity potential and leaf thickness. Plant leaf characteristics including water potential and osmotic water potential relationships have been studied with the device.

References

- Nutrient-doser, definition-1772: maximumyield.com, Retrieved 23 April, 2020
- An-introduction-to-automatic-dosing-in-hydroponics: blog.zipgrow.com, Retrieved 18 March, 2020
- Best-water-chiller-for-hydroponics: smartgardenguide.com, Retrieved 17 May, 2020
- Net-pot, definition-3094: maximumyield.com, Retrieved 16 January, 2020
- Max Clarke (13 March 2013). The Complete Guide to Building Your Own Greenhouse. Lulu.com. p. 53. ISBN 978-1-105-91367-9
- Lipp, Charles W., Practical Spray Technology: Fundamentals and Practice, 2012, ISBN 978-0-578-10090-6
- Mesh-pot, definition-3469: maximumyield.com, Retrieved 10 June, 2020

Permissions

All chapters in this book are published with permission under the Creative Commons Attribution Share Alike License or equivalent. Every chapter published in this book has been scrutinized by our experts. Their significance has been extensively debated. The topics covered herein carry significant information for a comprehensive understanding. They may even be implemented as practical applications or may be referred to as a beginning point for further studies.

We would like to thank the editorial team for lending their expertise to make the book truly unique. They have played a crucial role in the development of this book. Without their invaluable contributions this book wouldn't have been possible. They have made vital efforts to compile up to date information on the varied aspects of this subject to make this book a valuable addition to the collection of many professionals and students.

This book was conceptualized with the vision of imparting up-to-date and integrated information in this field. To ensure the same, a matchless editorial board was set up. Every individual on the board went through rigorous rounds of assessment to prove their worth. After which they invested a large part of their time researching and compiling the most relevant data for our readers.

The editorial board has been involved in producing this book since its inception. They have spent rigorous hours researching and exploring the diverse topics which have resulted in the successful publishing of this book. They have passed on their knowledge of decades through this book. To expedite this challenging task, the publisher supported the team at every step. A small team of assistant editors was also appointed to further simplify the editing procedure and attain best results for the readers.

Apart from the editorial board, the designing team has also invested a significant amount of their time in understanding the subject and creating the most relevant covers. They scrutinized every image to scout for the most suitable representation of the subject and create an appropriate cover for the book.

The publishing team has been an ardent support to the editorial, designing and production team. Their endless efforts to recruit the best for this project, has resulted in the accomplishment of this book. They are a veteran in the field of academics and their pool of knowledge is as vast as their experience in printing. Their expertise and guidance has proved useful at every step. Their uncompromising quality standards have made this book an exceptional effort. Their encouragement from time to time has been an inspiration for everyone.

The publisher and the editorial board hope that this book will prove to be a valuable piece of knowledge for students, practitioners and scholars across the globe.

Index

www.ingramcontent.com/pod-product-compliance
Lightning Source LLC
LaVergne TN
LVHW080616210926
848485LV00003B/167